AF693070

Trends and Challenges in Power System Stability and Control

Trends and Challenges in Power System Stability and Control

Guest Editors

Ke Xu

Yuan Liao

MDPI

Basel • Beijing • Wuhan • Barcelona • Belgrade • Novi Sad • Cluj • Manchester

Guest Editors

Ke Xu
Department of Electric Power Engineering
Nanjing Institute of Technology
Nanjing
China

Yuan Liao
Department of Electrical and Computer Engineering
University of Kentucky
Lexington, KY
USA

Editorial Office
MDPI AG
Grosspeteranlage 5
4052 Basel, Switzerland

This is a reprint of the Special Issue, published open access by the journal *Energies* (ISSN 1996-1073), freely accessible at: https://www.mdpi.com/journal/energies/special_issues/VSQB75217T.

For citation purposes, cite each article independently as indicated on the article page online and as indicated below:

Lastname, A.A.; Lastname, B.B. Article Title. *Journal Name* **Year**, *Volume Number*, Page Range.

ISBN 978-3-7258-5433-2 (Hbk)
ISBN 978-3-7258-5434-9 (PDF)
https://doi.org/10.3390/books978-3-7258-5434-9

© 2025 by the authors. Articles in this book are Open Access and distributed under the Creative Commons Attribution (CC BY) license. The book as a whole is distributed by MDPI under the terms and conditions of the Creative Commons Attribution-NonCommercial-NoDerivs (CC BY-NC-ND) license (https://creativecommons.org/licenses/by-nc-nd/4.0/).

Contents

energies

Article

Two-Stage Optimal Scheduling Based on the Meteorological Prediction of a Wind–Solar-Energy Storage System with Demand Response

Lu Wei [1,*], Yiyin Li [1], Boyu Xie [1], Ke Xu [2] and Gaojun Meng [2]

[1] Henan Meteorological Service Center, Zhengzhou 450003, China
[2] School of Electric Power Engineering, Nanjing Institute of Technology, Nanjing 211167, China; xukesam@njit.edu.cn (K.X.); gjun_m@njit.edu.cn (G.M.)
* Correspondence: j0000003013@njit.edu.cn

Abstract: With large-scale wind and solar power connected to the power grid, the randomness and volatility of its output have an increasingly serious adverse impact on power grid dispatching. Aiming at the system peak shaving problem caused by regional large-scale wind power photovoltaic grid connection, a new two-stage optimal scheduling model of wind solar energy storage system considering demand response is proposed. There is a need to comprehensively consider the power generation cost of various types of power sources, day-ahead load forecasting information, and other factors and plan the day-ahead output plan of the energy storage system with the minimum system operation cost as the optimization objective of day-ahead dispatching. The demand response strategy is introduced into the time-ahead optimal scheduling, and the optimization of the output value of the energy storage system in each period is studied with the goal of minimizing the system adjustment cost. The particle swarm optimization algorithm is used to solve the model, and the IEEE33 node system is used for an example simulation. The results show that using the demand response and the collaborative effect of the energy storage system can suppress the uncertainty of wind power and photovoltaic power, improve the utilization rate of the system, reduce the power generation cost of the system, and achieve significant comprehensive benefits.

Keywords: demand response; energy storage system; two-stage scheduling; particle swarm optimization algorithm

1. Introduction

With the large-scale integration of new energy into the grid, the instability and anti-peak regulation of its power generation output greatly increases the adjustment burden of the system [1]. This indirectly implies that vigorously improving the flexible regulation capacity of the power system, ensuring the balance of power supply and demand at different time scales, as well as the high-level absorption of new energy, is not only inherently demanded to speed up the construction of a new power system with new energy as the main body, but it is also an urgent requirement for the promotion of the realization of a carbon peak in 2030 and carbon neutralization in 2060 [2–4].

With the increase in the proportion of new energy access, the research on responding to the volatility and randomness of new energy by demand response has increased sharply [5,6]. In reference [7], a demand response scheduling strategy with multi-time scale rolling coordination was designed, which calls for a flexible load to track wind power rolling. In reference [8], a multi-objective optimization model of source–load coordination was established to maximize the absorption of wind power and minimize the operating cost of the system. In reference [9], a power system stochastic dispatching model considering large-scale wind power and demand response was established. In reference [10], a day-ahead optimization model of the power system was constructed by introducing time-sharing electricity price and interruptible load into a large-scale wind power system. In addition, the effects of two kinds of demand response resources to reduce the reverse peak regulation performance of wind power and the impact of intermittence on the power system were compared.

At present, there is almost no research on the analysis and modeling of demand response uncertainty [11]. Reference [12] analyzed the causes of the uncertainty of demand response resources and attributed the causes of the uncertainty to the uncertainty of the model and forecast. Reference [13] established the random response model of price type load. The uncertainty of price type load response is regarded as a random injection variable and introduced into the random interactive probabilistic power flow model of source load. Reference [14] artificially sets the range of uncertain demand response and reserves the uncertain response within this range so that the economy of the system is the highest. Reference [15] studied the incentive demand response and price demand response, and the scheduling cost models under uncertain response were established, respectively. The above literature provides a good theoretical basis for the uncertainty analysis of demand response. However, these studies focus on the modeling of the influencing factors of demand response uncertainty and analyze the relationship between influencing factors and demand response uncertainty, which cannot provide a more practical reference for power system dispatching.

In summary, this work constructs a two-stage optimal scheduling model of land-scape storage considering demand response, analyzes the load characteristics of the user side, and designs a DR model that integrates direct load control (DLC) and transferable load (TL) to make the load and new energy generation closer in timing. Subsequently, the scheduling optimization model of the day-ahead and the first two stages of the landscape storage system is established, and the demand response strategy is introduced in the pre-time scheduling phase to realize the comprehensive optimization of load demand distribution and energy storage output planning. Finally, the IEEE33 node system is taken as the simulation system to analyze the effect of the proposed model on the landscape ab-sorption ability of the system.

2. Load Characteristic Analysis and DR Modeling

2.1. User Side Load Characteristic Analysis

According to the way that the power system participates in the demand response, the customer-side load is divided into four categories [16]: (1) basic load: it belongs to uncontrollable load, which is fully responsive to the needs of users, and the system cannot change its energy use mode and time; (2) translational load: the power supply time of the load can be changed according to the plan, the load needs to be shifted as a whole, and the power absorption time spans multiple scheduling periods; (3) a load that can be reduced: it can withstand a certain interruption or power reduction, reduce the load running for a certain time, and reduce some or all of it according to supply and demand; and (4) transferable load: the electricity absorption in each time period can be adjusted flexibly, but the total load in the whole cycle should remain unchanged after transfer and before transfer.

2.2. Demand Response Modeling

2.2.1. DLC Model of Daily Electricity Absorption

The load of DLC is directly reduced by the load control device of the power company during the peak load, which is mainly aimed at residential or small commercial users and other domestic electricity users. The response characteristic of the participating load is that the load can be reduced, which belongs to peak-cutting resources, such as air conditioners, electric water heaters, and other loads with cold and hot storage capacity [17]. According to the dispatching needs and load distribution, multiple groups of users are involved in the DLC project, and the load reduction $F_{\mathrm{DLC}}(t)$ in the period t is as follows:

$$F_{\mathrm{DLC}}(t) = \sum_{i=1}^{I} \alpha_i(t) F_{\mathrm{DLC},i}(t) \tag{1}$$

where $F_{\mathrm{DLC},i}(t)$ represents the reducible load of i-th group user in the time t, $\alpha_i(t)$ represents 0–1 variable, $\alpha_i(t) = 1$ indicates that the load of i-th group is reduced, and $\alpha_i(t) = 0$ indicates that the load of i-th group is not reduced.

In order to encourage users to implement DLC, power suppliers will give electricity price compensation to participating users, and the rate of electricity price compensation that users can obtain when they participate in DLC is proportional to the degree of load reduction. Let the basic compensation rate of the DLC project be $\lambda_{0,\mathrm{DLC}}(\lambda_{0,\mathrm{DLC}} \in (0,1))$, then the actual electricity price compensation rate $\lambda_i(t)$ of i-th group users in the time t is

$$\lambda_i(t) = \lambda_{0,\mathrm{DLC}}(t)\left[\frac{\alpha_i(t) F_{\mathrm{DLC},i}(t)}{1.2 F_i(t)}\right] \tag{2}$$

where $F_i(t)$ represents the planned participation load of i-th group users in time t.

The load reduction compensation provided by the power supplier, that is, the DLC cost, is

$$C_{\mathrm{DLC}} = \sum_{i=1}^{I} \sum_{t=1}^{T} \lambda_i(t) P_1(t) \alpha_i(t) F_{\mathrm{DLC},i}(t) \Delta t \tag{3}$$

where $P_1(t)$ represents the domestic electricity price of the time t, T represents the number of scheduling periods, and Δt represents the scheduling step, with a value of 1 h.

2.2.2. TL Model of Industrial Power Absorption

TL entails users controlling themselves after receiving the system instruction signal; turn out or transfer to the load, mainly for large industrial users and other industrial power users; and that the response characteristic of the participating load is movable load. It is a kind of peak-cutting and valley-filling resource, and it is generally the industrial assembly line load that can be arranged for downtime production. There are K groups of users to participate in the TL project, and the transfer load $F_{\mathrm{TL}}(t)$ in the time t is

$$F_{\mathrm{TL}}(t) = \sum_{k=1}^{K} \rho(t) \alpha_k(t) F_{\mathrm{TL},k}(t) \tag{4}$$

where $F_{\mathrm{TL},k}(t)$ represents the transferable load of k-th group users in the time t, turning in is negative, turning out is positive; $\alpha_k(t)$ is 0–1 variable, $\alpha_k(t) = 1$ indicates that the load is selected, $\alpha_k(t) = 0$ indicates that the load is not selected, $\rho(t)$ represents the probability that the user obeys the power grid regulation and control, taking into account the uncertainty of user obedience, increases the load margin, which is more conducive to the stability of the system.

Similarly, the participating users of the DLC project are compensated according to the TL project, and the actual electricity price compensation rate $\lambda_k(t)$ and the TL cost C_{TL} are

$$\lambda_k(t) = \lambda_{0,\mathrm{TL}}(t)\left[\frac{|\rho(t)\alpha_k(t) F_{\mathrm{TL},k}(t)|}{1.2 F_k(t)}\right] \tag{5}$$

$$C_{\mathrm{TL}} = \sum_{k=1}^{K} \sum_{t=1}^{T} \lambda_k(t) P_2(t) |\rho(t) \alpha_k(t) F_{\mathrm{TL},k}(t)| \Delta t \tag{6}$$

where $\lambda_{0,\mathrm{TL}}$ represents the basic compensation rate of TL, $F_{k(t)}$ represents the planned participation load of k-th group users in time t, and $P_2(t)$ represents industrial electricity price in time t.

2.2.3. DR Comprehensive Model

In order to reduce the load more quickly and reliably during the peak hours of power absorption and to better improve the timing matching between load and new energy generation, the proposed DR comprehensive model takes into account both DLC and TL models. Then, the load $F_{\mathrm{DR}}(t)$ and the total call cost C_{DR} of DR after the implementation of DR in time t is

$$F_{\mathrm{DR}}(t) = F_{\mathrm{initial}}(t) - F_{\mathrm{DLC}}(t) - F_{\mathrm{TL}}(t) \tag{7}$$

$$C_{\mathrm{DR}} = C_{\mathrm{DLC}} + C_{\mathrm{TL}} \tag{8}$$

where $F_{\mathrm{initial}}(t)$ represents the initial load of time t.

2.3. User Satisfaction Degree

Users are important participants in the electricity market, and the implementation of DR will have an impact on their power comfort. Considering the scale of user participation, user satisfaction should be taken into account when implementing DR [18]. Customer satisfaction requires that the electricity demand is met in time, and the lower the load reduction or transfer, the higher the satisfaction.

$$\begin{cases} M_{\mathrm{DLC}} = \frac{1}{I}\sum\limits_{i=1}^{I}\sum\limits_{t=1}^{T}[1 - \frac{\alpha_i(t)F_{\mathrm{DLC},i}(t)}{F_i(t)}] \\ M_{\mathrm{TL}} = \frac{1}{K}\sum\limits_{k=1}^{K}\sum\limits_{t=1}^{T}[1 - \frac{|\rho(t)\alpha_k(t)F_{\mathrm{TL},k}(t)|}{F_k(t)}] \\ M_{\mathrm{user}} = \frac{1}{2}(M_{\mathrm{DLC}} + M_{\mathrm{TL}}) \end{cases} \tag{9}$$

where M_{DLC} and M_{TL} are the average user satisfaction of DLC and TL, respectively, and M_{user} is the comprehensive satisfaction of users.

3. Two-Stage Optimal Scheduling Model of Scenery Storage

3.1. Day-Ahead Scheduling Model

With the goal of minimizing the operating cost of the system, the output arrangement of the day-ahead energy storage system is established, and the objective function is as follows:

$$\min C_1 = C_{\mathrm{H}} + C_{\mathrm{W}} + C_{\mathrm{PV}} + C_{\mathrm{ESS}} \tag{10}$$

where C_1 represents the total operation cost of the whole dispatching cycle of the system, C_{H} represents the operation cost of the thermal power plant, C_{W} represents the cost of wind power generation, C_{PV} represents the cost of photovoltaic power generation, and C_{ESS} represents the operation cost of energy storage.

$$C_{\mathrm{H}} = \sum_{t=1}^{T_1}\sum_{i=1}^{n_{\mathrm{H}}} c_{\mathrm{H},i}p_{\mathrm{H},i}(t) \tag{11}$$

where T_1 represents the number of periods in a day-ahead dispatching cycle, n_{H} represents the number of thermal power units, $c_{\mathrm{H},i}$ represents the coal-fired cost of thermal power unit i, $p_{\mathrm{H},i}(t)$ represents the active power output of thermal power unit i at time t.

$$C_{\mathrm{W}} = \sum_{t=1}^{T_1}\sum_{i=1}^{n_{\mathrm{W}}} c_{\mathrm{W},i}p_{\mathrm{W},i}(t) \tag{12}$$

where n_{W} represents the number of wind farms, $c_{\mathrm{W},i}$ represents the power generation cost of wind farm i, $p_{\mathrm{W},i}(t)$ represents the active power output of wind farm i at time i.

$$C_{\mathrm{PV}} = \sum_{t=1}^{T_1}\sum_{i=1}^{n_{\mathrm{PV}}} c_{\mathrm{PV},i}p_{\mathrm{PV},i}(t) \tag{13}$$

where n_{PV} represents the number of photovoltaic power stations, $c_{\text{PV},i}$ represents the power generation cost of photovoltaic power station i, $p_{\text{PV},i}(t)$represents the active power output of photovoltaic power station i at time t.

$$C_{\text{ESS}} = \sum_{t=1}^{T_1} \sum_{i=1}^{n_{\text{ESS}}} c_{\text{ESS},i} [e_{ch,i}(t) p_{\text{ESS},i}(t)(1 - \eta_{ch,i}) + e_{dis,i}(t) p_{\text{ESS},i}(t)(1 - \eta_{dis,i})] \tag{14}$$

where n_{ESS} represents the number of energy storage power stations, $c_{\text{ESS},i}$ represents the operating cost of energy storage power station i, $p_{\text{ESS},i}(t)$ represents the active power output of energy storage power station i at time t, $e_{ch,i}(t)$ and $e_{dis,i}(t)$ represent the state variables of energy storage charge and discharge of energy storage power station i at time t, with values of 0 or 1. When the energy storage is charged, $e_{ch,i}(t) = 1$; when the energy storage is discharged, $e_{dis,i}(t) = 1$, and $e_{ch,i}(t) e_{dis,i}(t) = 0$; $\eta_{ch,i}$ and $\eta_{dis,i}$ are the charging efficiency and discharging efficiency of the energy storage power station i, respectively.

The constraints of the day-ahead scheduling model are as follows:

(1) Power balance constraints:

$$p_{load}(t) = \sum_{i=1}^{n_{\text{H}}} p_{\text{H},i}(t) + \sum_{i=1}^{n_{\text{W}}} p_{\text{W},i}(t) + \sum_{i=1}^{n_{\text{PV}}} p_{\text{PV},i}(t) + \sum_{i=1}^{n_{\text{ESS}}} p_{\text{ESS},i}(t) \tag{15}$$

where $p_{load}(t)$ represents the total load at time t.

(2) Upper and lower limits of power generation output:

$$\begin{cases} p_{\text{H},i}^{\min} \le p_{\text{H},i}(t) \le p_{\text{H},i}^{\max} \\ 0 \le p_{\text{W},i}(t) \le p_{\text{W},i}^{\max} \\ 0 \le p_{\text{PV},i}(t) \le p_{\text{PV},i}^{\max} \end{cases} \tag{16}$$

where $p_{\text{H},i}^{\min}$ and $p_{\text{H},i}^{\max}$ represent the minimum and maximum technical output of thermal power unit i; $p_{\text{W},i}^{\max}$ and $p_{\text{PV},i}^{\max}$ represent the maximum theoretical output of wind farm i and photovoltaic power station i, respectively.

(3) Climbing rate constraint:

$$\begin{cases} R_{\text{H},i}^{d} \le p_{\text{H},i}(t) - p_{\text{H},i}(t-1) \le R_{\text{H},i}^{u} \\ R_{\text{W},i}^{d} \le p_{\text{W},i}(t) - p_{\text{W},i}(t-1) \le R_{\text{W},i}^{u} \\ R_{\text{PV},i}^{d} \le p_{\text{PV},i}(t) - p_{\text{PV},i}(t-1) \le R_{\text{PV},i}^{u} \end{cases} \tag{17}$$

where $R_{\text{H},i}^{d}$ and $R_{\text{H},i}^{u}$ represent the lower limit and upper limit of the ramp rate of thermal power unit i, respectively; $R_{\text{W},i}^{d}$ and $R_{\text{W},i}^{u}$ represent the lower limit and upper limit of the ramp rate of wind farm i, respectively; $R_{\text{PV},i}^{d}$ and $R_{\text{PV},i}^{u}$ represent the lower limit and upper limit of the ramp rate of photovoltaic power station i, respectively.

(4) Energy storage active output constraints:

$$p_{\text{ESS},i}^{\min} \le p_{\text{ESS},i}(t) \le p_{\text{ESS},i}^{\max} \tag{18}$$

where $p_{\text{ESS},i}^{\min}$ and $p_{\text{ESS},i}^{\max}$ represent the lower limit value and upper limit value of the active power output of the energy storage power station i, respectively.

(5) Energy storage SOC constraints:

$$soc_{\min,i} \le soc_i(t) \le soc_{\max,i} \tag{19}$$

where $soc_i(t)$ represents the state of charge of energy storage power station i at time t; $soc_{\min,i}$ and $soc_{\max,i}$ represent the lower and upper limit of the state of charge of energy storage power station i, respectively.

In order to prevent the battery from being overcharged and overdischarged for a long time and prolong the energy storage life, this paper determines the priority and size of the output of each energy storage power station according to the SOC real-time status of the energy storage battery and reasonably arranges the current output plan of the energy storage system. The charge and discharge margin coefficients $C\%$ and $D\%$ are introduced to measure the dispatchability of each energy storage power station. The higher the value of $C\%$ and $D\%$, the greater the dispatchability and the priority output of the energy storage power station. The calculation equation is as follows:

$$C\% = \frac{soc_{\max,i} - soc_i(t)}{soc_{\max,i}} \times 100\% \tag{20}$$

$$D\% = \frac{soc_i(t) - soc_{\min,i}}{soc_{\min,i}} \times 100\% \tag{21}$$

3.2. Pre-Time Scheduling Model

The pre-time optimal scheduling model is based on the planned output curve of the day-ahead energy storage system and aims to minimize the system adjustment cost according to the user-side demand response. The adjustment cost mainly includes the adjustment cost of load participation demand response and the adjustment cost of modified energy storage output. The objective function is as follows:

$$\min C_2 = \sum_{t=1}^{T_2} [C_{\mathrm{DR}}(t) + \sum_{i=1}^{n_{\mathrm{ESS}}} w_{\mathrm{ESS},i}(t) \left| p_{\mathrm{ESS},i}(t) - p'_{\mathrm{ESS},i}(t) \right|] \tag{22}$$

where T_2 represents the number of periods in the pre-scheduling cycle, C_2 represents the adjustment cost of the whole scheduling cycle of the system, $C_{\mathrm{DR}}(t)$ represents the total DR call cost at time t, $w_{\mathrm{ESS},i}(t)$ represents the adjustment cost coefficient of the energy storage power station i, $p_{\mathrm{ESS},i}(t)$ represents the active power output of the energy storage power station i in the pre-time dispatching, $p'_{\mathrm{ESS},i}(t)$ represents the day-ahead dispatching plan value of the energy storage power station i active power output.

The constraints of the pre-time scheduling model are as follows:

(1)~(5) is similar to the equation in the day-ahead scheduling constraint.

(6) DR model-related constraints:

$$\alpha_i(t) F_{\mathrm{DLC},i}(t) \le F_i(t) \tag{23}$$

$$\left| \rho(t) \alpha_k(t) F_{\mathrm{TL},k}(t) \right| \le F_k(t) \tag{24}$$

(7) User satisfaction constraint:

$$M_{\mathrm{user}} \ge M_{\mathrm{user}}^{\min} \tag{25}$$

where $M_{\mathrm{user}}^{\min}$ represents the set minimum user satisfaction.

4. Evaluation Index

4.1. Peak-to-Valley Ratio

The greater the load peak-to-valley ratio, the more frequent the start-up and shutdown of the thermal power unit or the operation in the state of deep peak regulation, which affects the economic operation of the unit and interferes with the safety and stability of the system. Both DR and ESS can cut the peak, fill the valley, and reduce the ratio of peak to valley. The peak/valley ratio μ is defined as the ratio of the maximum load to the minimum load in a statistical period t. The smaller the value of μ, the better the effect of peak-cutting and valley-filling:

$$\mu = \frac{\max[F_{\mathrm{DR}}(t) - p_{\mathrm{ESS}}(t)]}{\min[F_{\mathrm{DR}}(t) - p_{\mathrm{ESS}}(t)]} \tag{26}$$

4.2. New Energy Absorption Rate

If the problem of trans-regional energy transmission is not considered, the new energy absorption is mainly related to the regulation performance of thermal power units (maximum and minimum technical output, power installation growth rate), load scale and peak–valley difference, energy storage scale, and maximum charging rate. The new energy absorption rate is defined as the ratio of the actual output of the new energy to its theoretical output [19]. The calculation equation is as follows:

$$\begin{cases} R_{\mathrm{W}} = \frac{Q_{\mathrm{W,act}} + Q_{\mathrm{W,ESS}}}{Q_{\mathrm{W,th}}} \\ R_{\mathrm{PV}} = \frac{Q_{\mathrm{PV,act}} + Q_{\mathrm{PV,ESS}}}{Q_{\mathrm{PV,th}}} \\ R_{\mathrm{new}} = \frac{Q_{\mathrm{new,act}} + Q_{\mathrm{new,ESS}}}{Q_{\mathrm{new,th}}} \end{cases} \tag{27}$$

where R_{W}, R_{PV}, and R_{new} represent wind, photovoltaic, and total new energy absorption rate, respectively; $Q_{\mathrm{W,act}}$, $Q_{\mathrm{PV,act}}$, and $Q_{\mathrm{new,act}}$ represent the direct output of wind, photovoltaic, and total new energy, respectively; $Q_{\mathrm{W,th}}$, $Q_{\mathrm{PV,th}}$, and $Q_{\mathrm{new,th}}$ represent the theoretical output of wind, photovoltaic, and total new energy, respectively; $Q_{\mathrm{W,ESS}}$, $Q_{\mathrm{PV,ESS}}$, and $Q_{\mathrm{new,ESS}}$ represent wind power, photovoltaic, and total new energy stored by ESS, respectively.

5. Case Analysis

5.1. Simulation Scenario Setting

Four kinds of system simulation scenarios are set to analyze the influence of the energy storage system and demand response on the peak-cutting, valley-filling, and landscape-absorbing ability of the system.

Case 1: The base scenario, which does not introduce energy storage systems and demand responses.

Case 2: The energy storage scenario, in which only the energy storage system is introduced, and four groups of energy storage are set up, each with a capacity of 50 MW·h. Other parameters are shown in the following section.

Case 3: The demand response scenario, which only introduces demand response. The proportion of planned participation load of DLC and TL projects to the total load is set to 0.05 and 0.08, respectively.

Case 4: The comprehensive scenario in which both demand response and energy storage systems are introduced. The specific parameters are the same as in Case 2.

5.2. Basic Data

This work is analyzed by taking the improved IEEE33 system as an example. The basic data of 10 thermal power units, DR and ESS, are shown in Tables 1–3, respectively. V_0 is the initial capacity of ESS.

The installed capacity of wind farms and photovoltaic power stations is 1500 and 500 MW, respectively. As shown in Figure 1, the photovoltaic output curve is set according to the photovoltaic power generation forecasting model, and the wind power output and load curve are set according to the historical forecast data. Assuming that the wind farm and photovoltaic power station are arranged to generate electricity according to the forecast, the costs of wind power and photovoltaic are 0.42 and 0.63 CNY (kW·h)$^{-1}$, respectively. In order to facilitate statistics, it is assumed that the user load changes of each group of DR projects are equal in one day, and the proportion of planned participation load to the total load of DLC and TL projects is 0.05 and 0.08, respectively.

Table 1. The basic data of thermal power unit.

Thermal Power	$p_{H,i}^{min}$/MW	$p_{H,i}^{max}$/MW	$R_{H,i}^{d}$/(MW·h^{-1})	$R_{H,i}^{u}$/(MW·h^{-1})	$c_{H,i}$/[yuan·(kW·h)$^{-1}$]
No. 1	150	400	−160	160	0.23
No. 2	120	300	−100	100	0.25
No. 3	120	300	−100	100	0.25
No. 4	100	300	−80	80	0.26
No. 5	100	300	−80	80	0.26
No. 6	100	300	−80	80	0.26
No. 7	50	200	−50	50	0.31
No. 8	50	200	−50	50	0.31
No. 9	50	200	−50	50	0.31
No. 10	50	200	−50	50	0.31

Table 2. The basic parameters of DR.

Period Type	Period/h	$P_1(t)$ /[yuan·(kW·h)$^{-1}$]	$P_2(t)$ /[yuan·(kW·h)$^{-1}$]	$\rho(t)$
Peak Hours	[10, 15] and [20, 23]	1.00	1.25	0.8
Valley Period	[0, 8]	0.30	0.40	1.0
Normal Period	else	0.55	0.80	0.9

Table 3. The basic parameters of ESS.

Energy Storage Parameters	ESS1	ESS2	ESS3	ESS4
V_0/(MW·h)	20	30	15	25
$soc_{\min,i}$	0.1	0.1	0.1	0.1
$soc_{\max,i}$	0.9	0.9	0.9	0.9
$\eta_{ch,I}$, $\eta_{dis,i}$	0.8	0.8	0.8	0.8
$c_{ESS,i}$/[yuan·(kW·h^{-1})]	0.08	0.08	0.08	0.08

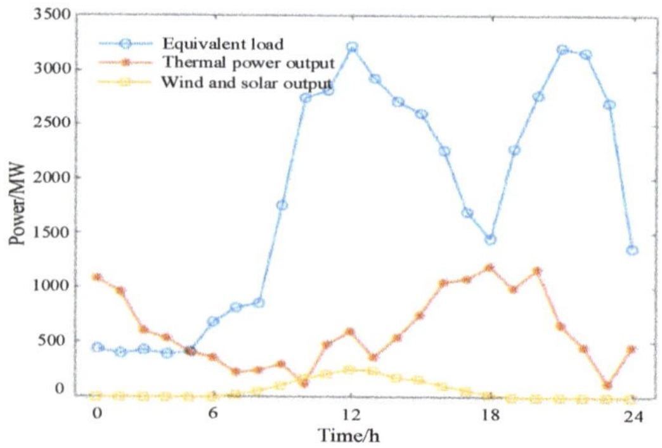

Figure 1. Load forecasting and theoretical output of wind power photovoltaic.

5.3. Scheduling Optimization Analysis

The hypothesis is to verify the effectiveness and practicability of the proposed model and to study the impact of DR and ESS on new energy absorption. The power generation dispatching of the system is divided into several scenarios for optimization analysis by MATLAB/Simulink 2023b simulation software.

Through the simulation experiment on scenarios 1–4, the corresponding equivalent load curve is shown in Figure 2, and the system optimization result is shown in Table 4. According to Figure 2, compared with the original load, the valley value of the equivalent load of the two-stage coordinated optimal scheduling during the valley period is higher than that of the equivalent load with only the energy storage scheduling scheme. It reaches its peak at about 13:00. At this time, the peak value of the equivalent load of the two-stage coordinated optimal scheduling is even smaller than that of the equivalent load, with only the energy storage scheduling scheme and other peak periods having similar effects. This proves that the two-stage coordinated optimal scheduling improves load curves and has a better effect on peak-cutting and valley-filling.

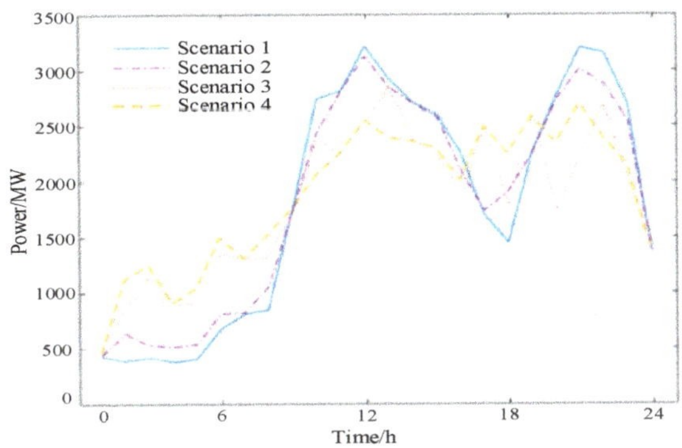

Figure 2. The curve of equivalent load.

Table 4. Optimization results under different scenarios.

Indicator	Scenario 1	Scenario 2	Scenario 3	Scenario 4
DR	No	No	Yes	Yes
ESS	No	Yes	No	Yes
S_{user}	1.00	1.00	0.74	0.76
μ	8.29	7.06	6.22	5.91
R_W	0.55	0.65	0.68	0.77
R_{PV}	0.97	0.96	0.98	0.99
R_{new}	0.63	0.68	0.70	0.81
C_1/10 thousand CNY	1174.25	1191.45	1192.11	1185.94
C_2/10 thousand CNY	0	58.23	59.76	65.12

According to Table 4, the proposed DR model effectively reduces the peak-to-valley ratio, which increases the absorption rate of new energy and ensures the smooth operation of the unit. After cooperating with ESS, the peak-to-valley ratio is further reduced, the level of new energy absorption is further improved, and the system is more stable. Combined with Figure 2, the reduction of the peak-to-valley ratio mainly improves the anti-peak regulation characteristics of wind energy, and the wind power absorption rate is significantly increased. The peak of photovoltaic output is consistent with the first peak of load, and the characteristic of anti-peak regulation is weak. After considering ESS, the discharge cost of ESS during the peak load period is much less than that of photovoltaic power generation. Affected by the lowest total cost of the system, the photovoltaic absorption rate may be slightly reduced, but the overall new energy absorption level still experiences a high degree of improvement.

As for the total operation cost, compared with scenario 1, the total cost of scenario 3 increased by CNY 178,600, and the total cost of scenario 4 increased by CNY 116,900; the new energy absorption level of scenarios 3 and 4 increased significantly, and the power generation cost was higher than that of thermal power, coupled with the call cost of ESS, so the total cost increased. The total cost of scenario 1 was low, but the peak and valley were relatively large, the system was not stable enough, and the safety and reliability were relatively low. For the problem of system adjustment cost, scenario 1 did not consider ESS and DR, so the value was 0; the adjustment cost of scenario 4 with comprehensive consideration of DR and ESS increases by CNY 53,600 compared with scenario 3, but the total operating cost of the system was reduced by CNY 61,700, so the overall economy is better. At the same time, scenario 4 had the lowest peak-to-valley ratio and the highest new energy absorption rate. To sum up, the comprehensive effect of the proposed model, namely scenario 4, is optimal.

The comparison of equivalent load and unit output under Scenario 4 is shown in Figure 3. The DR load change is shown in Figure 4. From 16:00 to 19:00, the actual peak period of new energy output is also the peak period of electricity load. From 0:00 to 5:00, the peak period of new energy output no longer corresponds to the trough period of load, which also makes the output of thermal power units relatively stable and improves the safety and stability of the system.

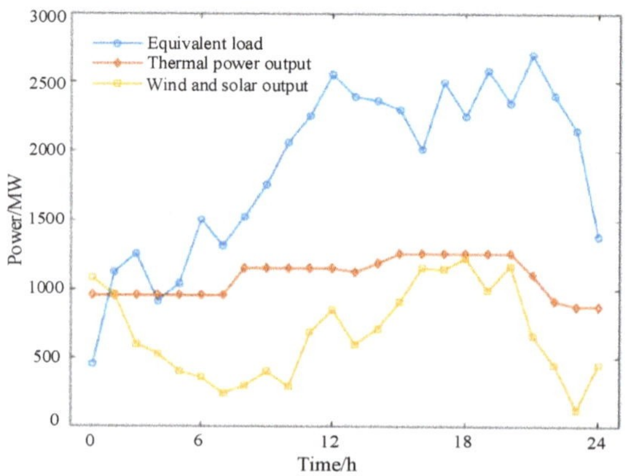

Figure 3. The comparison of equivalent load and unit output under scenario 4.

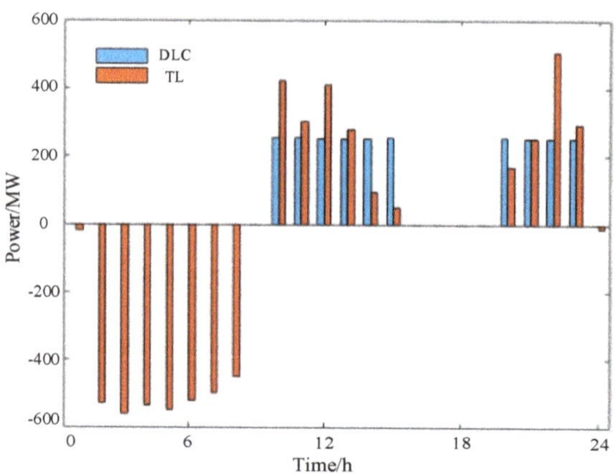

Figure 4. The demand response load changes under scenario 4.

From Figure 5a, when the load is at a low ebb, both the energy storage ESS1 and the energy storage ESS3 are in the charging state, and the charge of the energy storage ESS1 is less than that of the energy storage ESS3. According to Equation (20), during the whole valley period, the charge margin coefficient of energy storage ESS3 is larger than that of energy storage ESS1, so the scheduling priority of energy storage ESS3 is higher than that of energy storage ESS1. Therefore, when the system allocates power, the power allocated by the energy storage ESS3 is more than the energy storage ESS1.

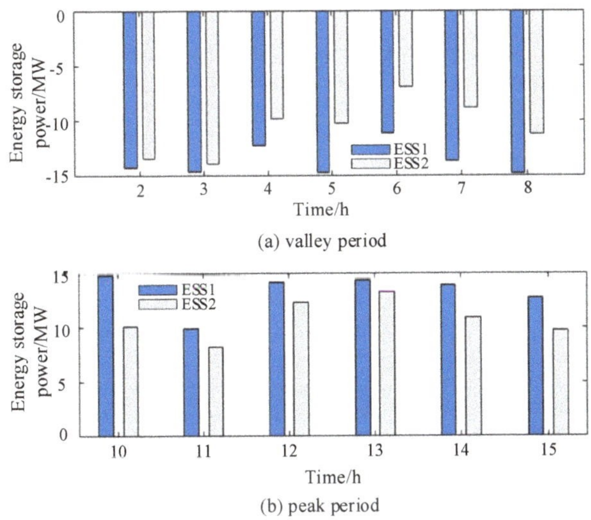

Figure 5. The comparison of energy storage output in different periods.

From Figure 5b, it can be seen that during the peak period of load, both energy storage ESS1 and energy storage ESS3 are in the state of discharge, and the discharge capacity of energy storage ESS1 is less than that of energy storage ESS3 during the peak period. During the whole valley period, the charging power of the energy storage ESS3 is greater than that of the energy storage ESS1, while in the peak period, the electricity of the energy storage ESS3 is larger than that of the energy storage ESS1. According to Equation (21), the discharge margin coefficient of energy storage ESS3 is larger than that of energy storage ESS1, so the scheduling priority of energy storage ESS3 is higher than that of energy storage ESS1. In general, the discharge of energy storage ESS3 is more than that of energy storage ESS1.

In order to reflect the superiority of the proposed DR model for different response groups over the conventional DR scheduling mode, generally only one response mode is considered. In this study, the new energy absorption rates of different proportions of DLC and TL project participation loads are compared, as shown in Figure 6. From Figure 6, the scheduling result considering different response modes is obviously better than that considering only one response mode, and the scheduling result is better with the increase of the proportion of the two response modes.

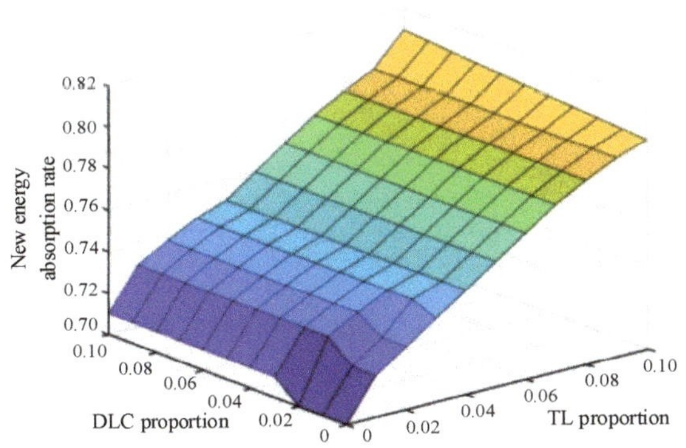

Figure 6. The new energy absorption rate under different demand response modes.

6. Conclusions

The reverse distribution of new energy output and load demand time is the main reason for a large number of abandonments of wind and light. In order to promote the large-scale grid connection of wind power and photovoltaic, a joint optimal scheduling model of wind and solar energy storage considering demand response is constructed based on the two-stage optimization theory. The results of numerical examples show the following:

1. The designed DR model effectively optimizes the demand-side load distribution. This, coupled with the call to the energy storage system, significantly improved the anti-peak regulation of new energy and increased the rate of new energy absorption.
2. The two-stage optimal scheduling model of landscape storage can optimize the output of the energy storage system (by modifying the day-ahead scheduling scheme), promote the energy storage system to participate in the optimal scheduling more reasonably, and improve the effect of peak-cutting and valley-filling.
3. The load peak-to-valley ratio is reduced through optimal dispatching; the output of the thermal power unit is more stable; the total cost is reduced; and the safety, reliability, and economic benefit of the system are improved.

The model proposed in this article has not yet taken into account the uncertainty and category of the load side, and the energy storage devices are relatively single. In subsequent research, the uncertainty of the load side and multiple types of energy storage devices can be added as research objects.

Author Contributions: Conceptualization, L.W.; Software, Y.L.; Validation, L.W.; Formal analysis, B.X. and K.X.; Resources, Y.L.; Writing—original draft, G.M.; Writing—review & editing, G.M.; Supervision, B.X. and K.X. All authors have read and agreed to the published version of the manuscript.

Funding: This research was funded by The Science and Technology Research Projects of Henan Provincial Department of Science and Technology, grant number 222103810093.

Data Availability Statement: The original contributions presented in the study are included in the article, further inquiries can be directed to the corresponding author.

Conflicts of Interest: The authors declare no conflict of interest.

References

1. Shu, Y.B.; Zhang, Z.G.; Guo, J.B.; Zhang, Z.L. Analysis of key factors of new energy absorption and research on solutions. *Proc. CSEE* **2017**, *37*, 1–8.
2. Tan, X.D.; Liu, J.; Xu, Z.C.; Yao, L.; Ji, G.Q.; Shan, B.G. Power supply and demand situation in the 14th five year plan under the "double carbon" goal. *Electr. Power* **2021**, *54*, 1–6.
3. Ju, L.W.; Yu, C.; Tan, Z.F. A two-stage scheduling optimization model and corresponding solving algorithm for power grid containing wind farm and energy storage system considering demand response. *Power Syst. Technol.* **2015**, *39*, 1287–1293.
4. Liu, W.Y.; Wen, J.; Xie, C.; Wang, W.Z.; Liang, C. Multi objective optimization method of power system source load coordination considering wind power absorption. *Proc. CSEE* **2015**, *35*, 1079–1088.
5. Stylianos, A.P. Integrated economic optimization of hybrid thermosolar concentrating system based on exact mathematical method. *Energies* **2022**, *15*, 7019.
6. Zang, H.; Ma, M.; Zhou, Y.; Xia, Q.; Sun, G.; Wei, Z. Robust optimal scheduling model for a 'wind power concentrating solar power-biomass' hybrid power plant in the electricity market. *Power Syst. Prot. Control* **2022**, *50*, 1–11.
7. Moretti, L.; Martelli, E.; Manzolini, G. An efficient robust optimization model for the unit commitment and dispatch of multi-energy systems and microgrids. *Appl. Energy* **2020**, *261*, 113859. [CrossRef]
8. Yu, D.; Sun, X.; Gao, B.T.; Xu, Q. Coordinated optimization model of wind power grid connection considering uncertain output of wind power. *Trans. China Electrotech. Soc.* **2016**, *31*, 34–41.
9. Mahmoud, N.; Saha, T.K.; Eghbal, M. Modelling demand response aggregator behavior in wind power offering strategies. *Appl. Energy* **2016**, *133*, 347–355. [CrossRef]
10. Mathieu, J.L.; Vayá, M.G.; Andersson, G. Uncertainty in the flexibility of aggregations of demand response resources. In Proceedings of the Industrial Electronics Society, IECON 2017, Vienna, Austria, 10–13 November 2017; pp. 8052–8057.
11. Zhou, J.; Yao, J.G.; Wang, K.; Shi, F.; Zeng, D. Continuous probabilistic power flow analysis method considering price type load response process. *Autom. Electr. Power Syst.* **2015**, *39*, 56–61.

12. Zhao, C.; Wang, J.; Watson, J.P.; Guan, Y. Multi-stage robust unit commitment considering wind and demand response uncertainties. *IEEE Trans. Power Syst.* **2012**, *280*, 2708–2717.
13. Niu, W.J.; Li, Y.; Wang, B.B. Modeling of demand response virtual power plant considering uncertainty. *Proc. CSEE* **2014**, *34*, 3630–3637.
14. Peng, C.H.; Liu, B.; Zuo, L.X.; Sun, H.J. Parallel multi-objective optimal scheduling of islanding microgrid considering classification demand response. *Power Syst. Prot. Control* **2019**, *47*, 60–68.
15. Li, P.; Xu, D.; Zhou, Z.; Lee, W.J.; Zhao, B. Stochastic optimal operation of microgrid based on chaotic binary particle swarm optimization. *IEEE Trans. Smart Grid* **2017**, *7*, 66–73. [CrossRef]
16. Liu, R.H.; Li, Z.L.; Yang, X.; Sun, G.P.; Li, L.H. The day ahead optimal scheduling of community integrated energy system considering user side flexible load. *Acta Energiae Solaris Sin.* **2019**, *40*, 2842–2850.
17. Qi, X.J.; Cheng, Q.; Wu, H.B.; Yang, S.H.; Li, Z.X. Influence of incentive demand response on distribution network operation reliability. *Trans. China Electrotech. Soc.* **2018**, *33*, 5319–5326.
18. Tang, W.; Gao, F. Recent optimal dispatching of household microgrid considering user satisfaction. *High Volt. Eng.* **2017**, *43*, 140–148.
19. Zhang, Z.Y.; Wang, W.L.; Zhang, G.; Wang, Z.W.; Ma, X.L.; Chu, Y.L. Evaluation method of new energy absorption capacity based on non time series model. *Autom. Electr. Power Syst.* **2019**, *43*, 24–30.

Disclaimer/Publisher's Note: The statements, opinions and data contained in all publications are solely those of the individual author(s) and contributor(s) and not of MDPI and/or the editor(s). MDPI and/or the editor(s) disclaim responsibility for any injury to people or property resulting from any ideas, methods, instructions or products referred to in the content.

energies

Article

Power Dispatching Strategy Considering the Health Status of Multi-Energy Conversion Equipment in Highway Power Supply Systems

Xianhong Hou [1], Jiao Wang [1], Shaoyong Guo [1] and Ketian Liu [2,*]

[1] CCCC Mechanical & Electrical Engineering Co., Ltd., Beijing 101300, China; houxh@ccccltd.cn (X.H.); wangjiao@ccccltd.cn (J.W.); guosy@ccccltd.cn (S.G.)
[2] School of Electric Power Engineering, Nanjing Institute of Technology, Nanjing 211167, China
* Correspondence: ktliu@njit.edu.cn

Abstract: In order to extend the service life of a highway power supply system and the level of new energy consumption, a power dispatching strategy considering the health status of multi-energy conversion equipment is proposed in this paper. Firstly, the energy and load forms of the highway power supply system are introduced, and the structure of the multi-energy conversion equipment, the topological structures of the DC–DC and DC–AC modules, and the operating characteristics are analyzed. Secondly, the module temperatures and output voltages are used as main parameters to establish the health indexes of DC–DC and DC–AC modules, and then the health index of the multi-energy conversion equipment is further calculated. Thirdly, the new energy consumption index is defined, and a multi-objective optimization model for power dispatching of highway power supply systems is established with the goal of improving the health index of multi-energy conversion equipment and the new energy consumption index. The case study shows that the power dispatching strategy in this paper can better control the temperature of each module, improve the health status of multi-energy conversion equipment, and have a high level of new energy consumption.

Keywords: highway power supply system; power dispatching strategy; multi-energy conversion equipment; health index; new energy consumption index

1. Introduction

Highways are important basic transportation facilities in the process of national economic construction and development. The power supply system plays a vital role in maintaining the normal operation of the highway system. Highway power supply systems are generally located in remote areas with a large power supply radius and weak connection with the power grid. On the other hand, the highway power supply systems have gradually become polymorphic in terms of energy supply and load type, with polymorphic energy and load access such as wind, solar, electricity storage, heat, and hydrogen. Especially with the popularization of new energy vehicles, the charging load of electric vehicles has grown rapidly, and the peak-to-valley difference of the power supply system has increased day by day. The highway power supply system has polymorphic power load, changeable external climate, and untimely maintenance, and the problem of power supply reliability is becoming increasingly prominent.

Currently, there are few studies on the power dispatching of highway power supply systems. The main studies focus on multi-energy complementary considering economic dispatching and new energy consumption. In terms of economic dispatching, reference [1] proposes a comprehensive economic model of the multi-microgrid for optimizing the power dispatching, and the source network load storage is taken into account. For the full consumption of photovoltaic (PV) power, reference [2] uses HOMER Grid software simulation to propose the economic benefit analysis method of PV plus battery energy storage systems (BESS) applied to the behind-the-meter (BTM) market, which is to consider the effective use of renewable energy. Reference [3] comprehensively considers the total operating cost of the system and the fluctuation of total power output and optimizes the dispatching of the wind power, PV power, and hydropower complementary systems. Reference [4] studies the economic efficiency of coal in the optimal dispatching of wind power, hydropower, and thermal power based on the randomness of wind power and load. Reference [5] considers the peak and frequency regulation of the system to establish a wind power, hydropower, and thermal power optimal dispatching model. Reference [6] studies the optimal dispatching of hydropower and wind power coordination based on influencing factors such as wind power, electricity prices, and water resources reserves. Reference [7] considers the charging and discharging power performance of an energy storage system to establish a wind power, hydropower, and energy storage system optimization dispatching model.

In terms of new energy consumption, in reference [8], the new operational conditions of conventional generators are taken into account for the safety and economics of the power system with a high proportion of renewable energy. Reference [9] establishes an optimal dispatching model for wind, solar, water, thermal, and energy storage. The objective function is to minimize the comprehensive costs of power generation, pollutant control, and renewable energy curtailment. The results show that the model can reduce the curtailment rate of renewable energy and the system operating costs and effectively smooth out power fluctuations. Reference [10] establishes a wind–solar–thermal–storage optimization dispatching model based on the optimal wind-solar energy curtailment rate, indicating that reasonable energy curtailment is conducive to improving the overall economic benefits of the system. Reference [11] establishes a wind–storage–hydro–thermal optimization dispatching model based on the load mean deviation method and multi-objective optimization solution strategy, effectively exerting the role of multi-energy complementarity and improving the consumption of clean energy. Reference [12] establishes a wind–hydro–thermal joint dispatching model by optimizing carbon emissions through electricity substitution and improving the consumption of new energy and environmental protection. Reference [13] proposes a multi-energy complementary optimization dispatching model for wind power, PV power, hydropower, thermal power, and energy storage with consideration of the impacts of the peak load regulation initiative of thermal power units. The research results show that the model proposed in [13] is helpful to improve the consumption of new energy. Reference [14] considers the randomness and volatility of renewable energy and quantifies the power balance problem of power system with high proportion renewables. Reference [15] focuses on power systems a with high proportion of renewable energy and proposes key measures to solve the problem of renewable energy consumption from multiple aspects, including the construction of complementary power sources, flexibility transformation of thermal power units, and demand-side response.

In summary, current power dispatching strategies rarely pay attention to the operating status of key equipment in the power grid. The health index method originated from the assessment of human health status and is also used to indicate the expected performance of the evaluated object [16]. It has been applied to measure and indicate the comprehensive performance of objects in many fields, such as spacecraft, aircraft engines, rocket engines, bridges, and other physical systems [17–19]. The concept of electric equipment health was first proposed by British scholar D. Hughess in 2003. After years of development, electric equipment health has gradually become a concept and expanded to multiple areas of power supply reliability. Most of the health analyses in the power industry use evaluation methods based on expert experiences [20–24].

Multi-energy conversion equipment is the core power electronic equipment in highway power supply systems. Current research ignores the impacts of the converter operating status, especially temperature, on the health status of the converter. Therefore, this paper takes a highway power supply system in southwest China as the research background and the multi-energy conversion equipment, the core equipment of the power supply system, as the research object. It analyzes the factors affecting and evaluating the reliable operation of multi-energy conversion equipment, establishes the health index of the multi-energy conversion equipment and the multi-objective optimization model of power dispatching based on the health index, and extends the service life of the multi-energy conversion equipment on the basis of meeting the energy demand of the highway power system.

The remainder of this paper is structured as follows: Section 2 introduces the basic functions and structure of the multi-energy conversion equipment of the highway power supply system, the DC–DC and DC–AC module topologies, analyzes the fever conditions of the modules, and proposes a method for calculating the health index of the multi-energy conversion equipment based on the temperatures and voltages. Section 3 establishes a multi-objective power control strategy that comprehensively considers the health status of multi-energy conversion equipment and the consumption level of new energy. The constraints are the module power limitations, state of charge (SOC) of energy storage system constraints, and hydrogen production constraints. Section 4 is the case study, which verifies and analyzes the power dispatching strategy of the highway power supply system proposed in this paper. The result shows that the power dispatching strategy proposed in this paper can improve the health index of multi-energy conversion equipment while meeting the power demand of highway service areas and, at the same time, has a higher level of new energy consumption.

2. Health Status Analysis of Multi-Energy Conversion Equipment

2.1. Multi-Energy Conversion Equipment of Highway Power Supply Systems

Multi-energy conversion equipment is manufactured by BJ-NEGO Automation Technology Co., Ltd. in Beijing, China. As the core electric device of the highway power supply system, the multi-energy conversion equipment takes power conversion as its main task and is supported by the power system to achieve multi-energy conversion and coordinated power dispatching based on wind power, PV power, power grid electricity power, and an energy storage system. The internal modules of the multi-energy conversion equipment are divided into two categories according to the circuit topology: DC–DC modules and a DC–AC module. The multi-energy conversion equipment adopts a rack mode, and multiple DC–DC modules are inserted into the racks vertically or horizontally to ensure its heat dissipation capacity and neatness. Among them, the DC–DC modules are connected to the 1500 V high-voltage DC bus, which is responsible for the energy access and control of wind power, PV power, energy storage, fuel cells, and DC loads. Wind power and PV power are the main power sources for highway power supply systems. The energy storage system is responsible for the stability of DC and cooperates with system scheduling to implement multi-objective control strategies. The bidirectional DC–AC module is connected to the 690 V AC bus, which is responsible for the DC-to-AC inversion, connected to the external power grid, and supplies power to each energy-consuming unit in the highway power supply system. The structure of the highway power supply system with multi-energy conversion equipment is shown in Figure 1.

The rated capacity of a single DC–DC module in the multi-energy conversion equipment is 270 kW. It adopts a three-phase interleaved parallel three-level buck–boost converter. The topology is shown in Figure 2a. Each module consists of 12 insulate-gate bipolar transistors (IGBT). The power consumption of a single IGBT is 321.2 W, and the total power consumption of IGBTs is 3854.4 W. According to the simulation condition of ambient temperature (air inlet) of 55 °C, the maximum temperature simulation result of IGBT is 122.5 °C. The rated capacity of the bidirectional DC–AC module is 1.725 MW. It adopts an I neutral point clamped (I-NPC) three-level inverter. The topology is shown in Figure 2b. The module consists of 12 IGBTs. The total module loss is 6208.8 W and the maximum temperature is 108.9 °C.

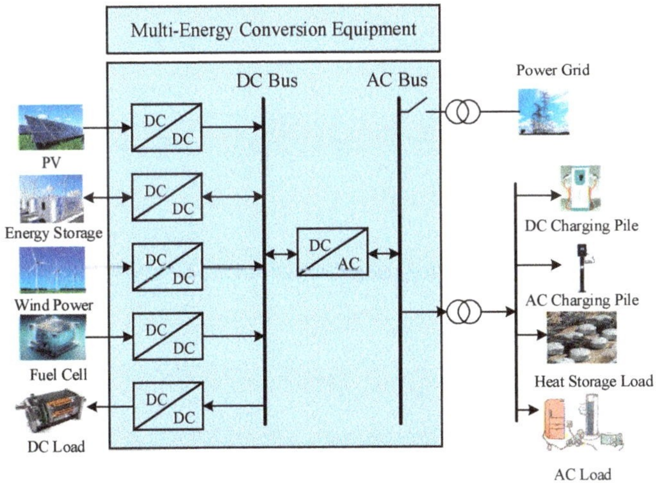

Figure 1. The structure of the highway power supply system.

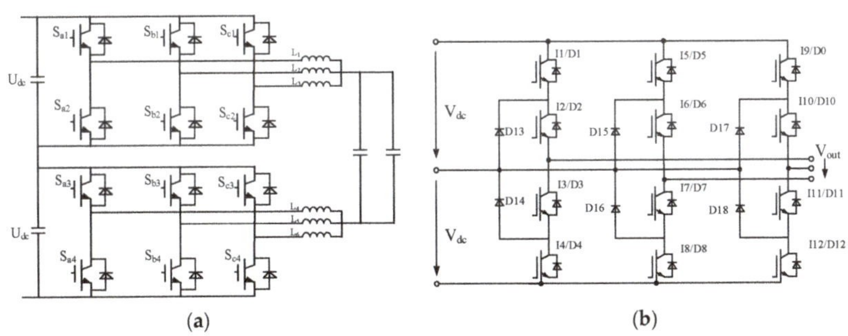

Figure 2. The topologies of DC–DC and DC–AC modules. (**a**) DC–DC module topology; (**b**) DC–AC module topology.

2.2. Health Index of Multi-Energy Conversion Equipment Based on Operating Status

In the early stage of highway power supply system operation, the historical operation and maintenance data of multi-energy conversion equipment are lacking, and it is impossible to establish the outage model of each component like the traditional reliability analysis method. Therefore, this paper evaluates the operating status of multi-energy conversion equipment based on the operating parameters such as the temperature of each module of multi-energy conversion equipment and the AC and DC bus voltages and establishes the health index of multi-energy conversion equipment.

The closer the module's operating temperature is to the lower limit of the normal operating temperature, the better the module's operating status is and the greater the adjustable power of the module is. The closer the module temperature is to the upper limit of the normal operating temperature, the worse the module's operating status is. When it exceeds the upper limit, it means that the module may be abnormal. The module health index based on temperature is as follows:

$$h_{\mathrm{T}}=\begin{cases}1, & T\leq T_{\min}\\ 1-\frac{T-T_{\min}}{T_{\max}-T_{\min}}, & T_{\min}<T<T_{\max}\\ 0, & T\geq T_{\max}\end{cases} \tag{1}$$

where h_T is the health index of DC–DC module or DC-AC module based on temperature; T is the DC–DC or DC-AC module operating temperature measured; and $[T_{min}, T_{max}]$ is the module normal operating temperature range.

The DC and AC voltages output by DC–DC and DC–AC modules should be within a reasonable range. When the output voltage deviates from the normal operating voltage range, it indicates that the operating status of the module may deteriorate. When the output voltage exceeds the module operating voltage limit, it means that the module may be abnormal. The module health index based on the module AC or DC voltage is as follows:

$$h_V = \begin{cases} 0, & V \le V_{min} \\ 1 - \frac{V_a - V}{V_a - V_{min}}, & V_{min} < V < V_a \\ 1, & V_a < V < V_b \\ 1 - \frac{V - V_b}{V_{max} - V_b}, & V_b < V < V_{max} \\ 0, & V \ge V_{max} \end{cases} \tag{2}$$

where V is the operating output voltage of the DC–DC or DC–AC module; $[V_a, V_b]$ is the normal range of the module output voltage; and $[V_{min}, V_{max}]$ is the allowable operating voltage of the module.

The operating parameters of the DC–DC module are the IGBT temperature and the DC bus output voltage. The health index of a single DC–DC module is calculated as follows:

$$h_{DC} = \alpha h_T + \beta h_V \tag{3}$$

where α and β are weight coefficients based on temperature and DC voltage health indicators, respectively.

The main operating parameters of the DC–AC module are IGBT temperature, AC side inductor temperature, and AC output voltage. The health index of a single DC–AC module is calculated as follows:

$$h_{AC} = \alpha_1 h_{T1} + \alpha_2 h_{T2} + \beta h_V \tag{4}$$

The multi-energy conversion equipment consists of multiple DC–DC modules and one DC–AC module. The comprehensive health index of a single multi-energy conversion equipment is calculated as follows:

$$H = \sum_{i=1}^{N} \alpha_i h_{DC,i} + \beta h_{AC} \tag{5}$$

where N is the number of DC–DC modules in the multi-energy conversion equipment.

3. Multi-Objective Optimization Model for Power Dispatching Strategy of Highway Power Supply System Considering Equipment Health and New Energy Consumption

The power dispatching of highway power supply systems must not only consider the health status of multi-energy conversion equipment but also take into account the power supply system's consumption of new energy, make full use of the new energy power generation in the power supply system, improve the energy self-consistency rate of the power supply system, and reduce the cost of purchasing electricity from the external power grid. Therefore, the establishment of the new energy consumption index is as follows [15]:

$$L = 1 - \left| \frac{S \cdot P - P_s}{P_s} \right| \tag{6}$$

where L is the new energy consumption index; S is the startup status of the new energy unit, 1 indicates startup and 0 indicates shutdown; P is the actual output power of the new energy unit; and P_s is the target power or predicted power of the new energy unit.

3.1. Objective Function

On the basis of considering the conventional control objectives of highway power supply systems, the health status information of multi-energy conversion equipment and the new energy consumption level are integrated to establish a multi-objective optimization model for power dispatching of highway power supply systems. The objective function is as follows:

$$J = \max\left\{ a\sum_{j=1}^{C}\sum_{i=1}^{M} H_i^j + b\sum_{j=1}^{C}\sum_{k=1}^{K} S_k^j L_k^j \right\} \tag{7}$$

where C is the number of optimized cycles; M is the number of multi-energy conversion equipment; and K is the number of new energy unit installed.

3.2. Constraints

1. The power constraints of each new energy unit are as follows:

$$P_i^{j,\min} \le P_i^j \le P_i^{j,\max} \tag{8}$$

where, $P_i^{j,\min}$ is the lower limit of the output power of the new energy unit in the jth control period, which is generally determined by the minimum operating power of the new energy unit; and $P_i^{j,\max}$ is the upper limit of the output power of the new energy unit in the jth control period, which can be equivalent to its predicted power.

2. In order to cope with the power demand of the highway power supply system when the external grid power supply is lost, the SOC constraint of the energy storage system at any time is as follows:

$$SOC_{\min} \le SOC \le SOC_{\max} \tag{9}$$

During the entire dispatching cycle, the energy storage system may be in a charging state or a discharging state. The energy storage system SOC calculation formula is as follows:

$$\begin{cases} SOC_t = SOC_{t-1} + \eta P_c \cdot t \\ SOC_t = SOC_{t-1} - \eta P_d \cdot t \end{cases} \tag{10}$$

where, $SOC_{t\text{-}1}$ is the SOC of the energy storage system after the end of the previous dispatching cycle; P_c and P_d are the charging and discharging power of the energy storage battery, respectively; and η is the charging and discharging efficiency of the energy storage power station.

3. The module temperature is mainly affected by the module power. Simulation analysis shows that the module temperature is linearly related to the power. Therefore, the relationship between the module temperature and power can be approximately expressed as follows:

$$T = T_0 + k(P - P_0) \tag{11}$$

where, k is the proportional coefficient, which can be obtained through experiments; T_0 is the initial temperature of the module; and P_0 is the initial power.

4. The module output voltage is also related to the power of the module. When the power increases, the module output voltage decreases. When the power decreases, the module output voltage increases. The relationship between voltage and power is approximately expressed as follows:

$$V = V_0 + k(P_0 - P) \tag{12}$$

where V_0 is the voltage corresponding to the module power P_0.

5. The power constraints of the DC–DC module and the DC–AC module. Considering the loss of the DC–DC module, the power of the DC–AC module is as follows:

$$P_{AC} = P_{DG} - P_{DL} - P_{Dl} \tag{13}$$

where P_{AC} is the DC–AC module power from DC side to AC side; P_{DG} is the total generation power of new energy; P_{DL} is the total load power of DC side; and P_{Dl} is the total power loss of DC–DC modules.

4. Case Analysis

In some highway power supply systems, new energy power generation includes wind power and PV power. The energy storage system is mainly used to supply power to the system in an emergency state, and to smooth the fluctuation of new energy power and improve the new energy consumption capacity. The DC load is the hydrogen production load. It starts when the generation of new energy is high, and corresponding production tasks are formulated every day. The PV power installed capacity is 900 kW, which is connected with multi-energy conversion equipment through four DC–DC modules. The wind power installed capacity is 400 kW, which is connected with multi-energy conversion equipment through two DC–DC modules. The level 1 load (non-interruptible load) in the power supply system is 50 kW. The energy storage system must be able to provide continuous power supply for 12 h. The energy storage system is configured to be 1000 kWh. Ignoring the charge and discharge loss of the energy storage system and the change of the module voltage, the minimum SOC constraint of the energy storage system is 60%~90%, the power consumption of hydrogen production is 200 kWh, the initial temperature of the DC–DC module is 90 °C, and the corresponding power is 150 kW. The proportional coefficient k is 0.1333. Only the effect of temperature on the health of multi-energy conversion equipment is considered, assuming that the normal temperature range of each module is [85, 105]. Figure 3 shows the load curve, PV power, and wind power curves of the highway power supply system.

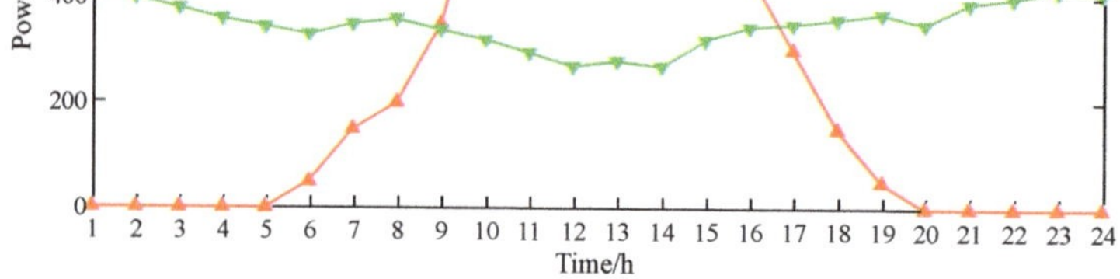

Figure 3. The load, PV power and wind power of highway power supply system.

In Figure 3, from 10:00 to 15:00, the PV power output is at a high level, and combined with the wind power output, it can meet the load demand of the highway power supply system. At other times, electricity needs to be imported from the external power grid. Figure 4 compares the PV power and wind power outputs before and after power dispatching and shows the external power grid input power, the energy storage system charging and discharging power, and the hydrogen production power after power dispatching.

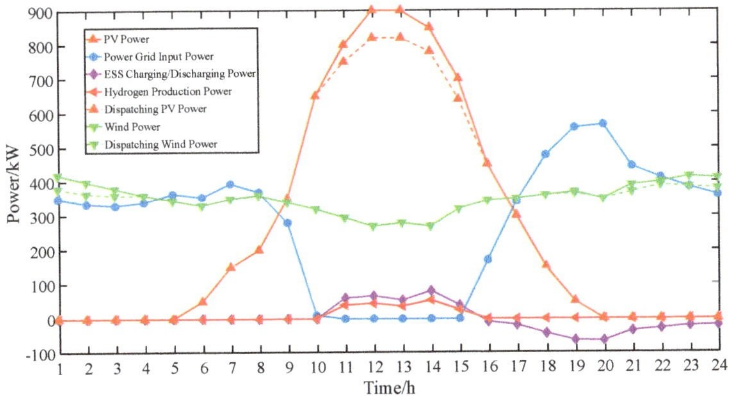

Figure 4. Dispatching of PV power, wind power, external power grid input power, energy storage and hydrogen production.

It can be seen in Figure 4 that when the PV power and wind power powers are high, in order to reduce the temperatures of the DC–DC modules, the dispatching strategy appropriately reduces the output power of PV power and wind power. From 11:00 to 15:00, during the PV power at high power period, in order to make full use of new energy, the energy storage system charges and the hydrogen production produces hydrogen. From 17:00 to 24:00, the PV power is low, the PV power is not controlled, and in order to reduce the amount of electricity imported from the external power grid, the energy storage system discharges until the SOC reaches the minimum limit of 60%.

Figure 5 compares the temperatures of the PV power and wind power DC–DC modules before and after power dispatching. It can be seen that after the implementation of power dispatching, the temperature of the PV power module is significantly reduced during the peak period and the temperature of the wind power module reduces during the entire dispatching cycle. The maximum temperature of the photovoltaic module drops from 105 °C to 102 °C, and the maximum temperature of the wind power module drops from 103 °C to 88 °C.

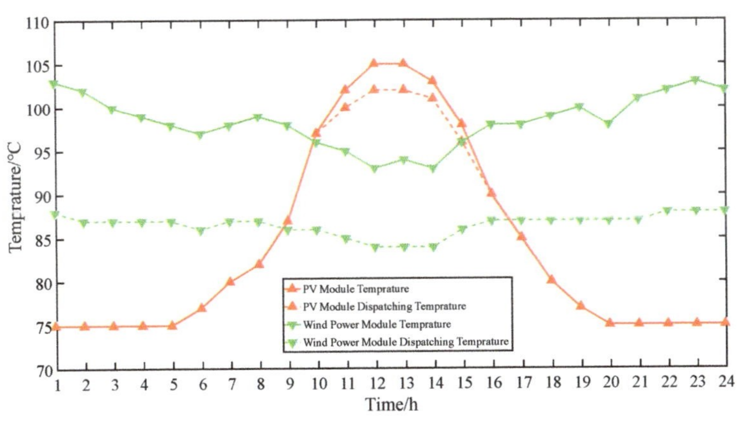

Figure 5. Temperature comparisons of different modules before and after power dispatching.

Assume that new energy is fully consumed before power dispatching. In order to further verify the effectiveness of the power dispatching strategy in this paper, the health index of multi-energy conversion equipment, the new energy consumption index, and the objective function value of the power dispatching optimization strategy before and after power dispatching are calculated, as shown in Figure 6.

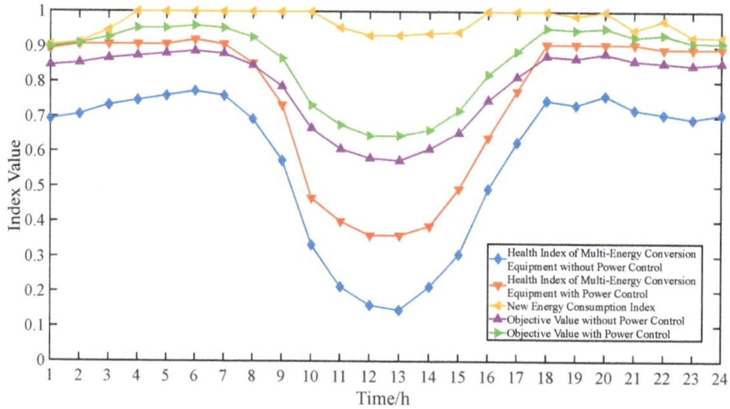

Figure 6. Index values comparisons before and after power dispatching.

It can be seen that after adopting the power dispatching strategy, the health index of multi-energy conversion equipment has been significantly improved, while the consumption level of new energy is still at a high level. The abandoned PV and wind energy is 530 kWh, and the objective function value is improved throughout the entire power dispatching period.

5. Conclusions

The main novelties of this paper are establishing the health index of multi-energy conversion equipment based on the temperature and output voltage of each module and proposing a power dispatching optimization model based on the health index of multi-energy conversion equipment and the new energy consumption index. The conclusions and suggestions are as follows:

(1) In multi-energy conversion equipment, the module temperature and power have a certain relationship, and temperature is an important parameter affecting the multi-energy conversion equipment health status;

(2) The health index based on temperature and output voltage can describe the operating status of multi-energy conversion equipment to a certain extent;

(3) The power dispatching strategy based on equipment health index and new energy consumption index in this paper can reduce the module temperature during the peak period of new energy power while taking into account the consumption of new energy and can delay the aging process of multi-energy conversion equipment and improve the reliability of the power supply system. In addition, since wind power lasts for a long time and has a large heat accumulation, wind power dispatching should be given priority.

Author Contributions: Conceptualization, X.H. and J.W.; methodology, X.H. and J.W.; software, S.G.; validation, X.H.; formal analysis, J.W.; resources, J.W.; writing—original draft preparation, K.L.; writing—review and editing, K.L.; supervision, X.H. All authors have read and agreed to the published version of the manuscript.

Funding: This work was supported by the National Key R&D Program of China (2021YFB2601404).

Data Availability Statement: The original contributions presented in the study are included in the article. Further inquiries can be directed to the corresponding author.

Conflicts of Interest: Authors Xianhong Hou, Jiao Wang and Shaoyong Guo were employed by the company CCCC Mechanical & Electrical Engineering Co., Ltd. The remaining authors declare that the research was conducted in the absence of any commercial or financial relationships that could be construed as a potential conflict of interest.

References

1. Du, X.; Wang, L.; Zhao, J.; He, Y.; Sun, K. Power Dispatching of Multi-Microgrid Based on Improved CS Aiming at Economic Optimization on Source-Network-Load-Storage. *Electronics* **2022**, *11*, 2742. [CrossRef]
2. Peng, C.Y.; Kuo, C.C.; Tsai, C.T. Optimal Configuration with Capacity Analysis of PV-Plus-BESS for Behind-the-Meter Application. *Appl. Sci.* **2021**, *11*, 7851. [CrossRef]
3. Zhang, Q.W.; Wang, X.L.; Li, Y. Optimal Scheduling of Hybrid Wind-Photovoltaic-Hydro-Pumped Storage Power System. *Power Energy* **2017**, *38*, 581–586.
4. Shi, L.; Yao, L.; Wang, R. Modeling and Solutions of Coordinated Economic Dispatch with Wind-Hydro-Thermal Complex Power Source Structure. *IET Renew. Power Gener.* **2016**, *11*, 262–270. [CrossRef]
5. Ge, X.L.; Jin, Y.; Xia, S.; Fu, Y. A Coordinated Optimization Scheduling of Wind-hydro-thermal Power System Based on Requirement of Peak Load and Frequency Regulation. *Power Syst. Technol.* **2019**, *43*, 3918–3925.
6. Chen, Q.; Yang, X.; He, G.Q. Optimal scheduling system for wind farm and hydro power plant coordinating operation. *Appl. Energy Symp. Forum* **2018**, *145*, 277–282. [CrossRef]
7. Zhao, S.Q.; Liu, D.Z.; Xie, Y.Q.; Liu, J.; Hu, Y.Q. Scheduling Strategy of Energy Storage in Wind-Solar-Battery Hybrid Power System Based on Dependent-Chance Goal Programming. *Autom. Electr. Power Syst.* **2015**, *39*, 30–36.
8. An, L.; Wang, M.B.; Qi, X.; Li, F.H.; Qu, J.H. Optimal Dispatching of Multi-Power Sources Containing Wind/Photovoltaic/Thermal/Hydro-Pumped and Battery Storage. *Renew. Energy Resour.* **2018**, *36*, 1492–1498.
9. Zhang, G.B.; Chen, Y.; Zhang, J.H.; Tang, N.N. Research on Optimization of Day-Ahead Dispatching of Wind Power-Photovoltaic-Hydropower-Thermal Power-Pumped Storage Combined Power Generation System. *Acta Energiae Solaris Sin.* **2020**, *41*, 80–85.
10. Ye, Z.; Li, X.Q.; Jang, F.; Chen, L.; Wang, Y.L.; Dai, S.F. Hierarchical Optimization Economic Dispatching of Combined Wind-PV-thermal-energy Storage System Considering the Optimal Energy Abandonment Rate. *Powre Syst. Technol.* **2021**, *45*, 2270–2279.
11. Wang, K.Y.; Luo, X.J.; Jia, R.; Zhou, C.W. Short-term Coordinated Scheduling of Wind-pumped-hydro-thermal Power System With Multi-energy Complementarities. *Powre Syst. Technol.* **2020**, *44*, 3631–3640.
12. Wang, Y.M.; Zhao, M.Z.; Chang, J.X.; Wang, X.B.; Tian, Y.Y. Study on the Combined Operation of A Hydro Thermal-Wind Hybrid Power System Based on Hydro-Wind Power Compensating Principles. *Energy Convers. Manag.* **2019**, *194*, 94–111. [CrossRef]
13. Li, T.; Li, Z.W.; Yang, J.Y.; Cui, D.; Wang, Z.H.; Ma, K.; Hu, W. Coordination and Optimal Scheduling of Multi-energy Complementary System Considering Peak Regulation Initiative. *Powre Syst. Technol.* **2020**, *44*, 3623–3630.
14. Li, M.J.; Chen, G.P.; Dong, C.; Liang, Z.F.; Wang, W.S.; Fan, G.F. Research on Power Balance of High Proportion Renewable Energy System. *Powre Syst. Technol.* **2019**, *43*, 3979–3986.
15. Shu, Y.B.; Zhang, Z.G.; Guo, J.B.; Zhang, Z.L. Study on Key Factors and Solution of Renewable Energy Accommodation. *Proc. CSEE* **2017**, *37*, 1–8.
16. Veillard, J.; Champagne, F.; Klazinga, N.; Kazandjian, V.; Arah, O.A.; Guisset, A.L. A performance assessment framework for hospitals: The WHO regional office for Europe PATH project. *Int. J. Qual. Health Care J. Int. Soc. Qual. Health Care* **2005**, *17*, 487–496. [CrossRef]
17. Yang, F.; Habibullah, M.S.; Zhang, T.; Xu, Z.; Lim, P.; Nadarajan, S. Health index-based prognostics for remaining useful life predictions in electrical machines. *IEEE Trans. Ind. Electron.* **2016**, *63*, 2633–2644. [CrossRef]
18. Liu, K.; Gebraeel, N.; Shi, J. A Data-Level Fusion Model for Developing Composite Health Indices for Degradation Modeling and Prognostic Analysis. *Autom. Sci. Eng. IEEE Trans.* **2013**, *10*, 652–664. [CrossRef]
19. Xu, L.; Xu, J.P. Integrated system health management-based progressive diagnosis for space avionics. *IEEE Trans. Aerosp. Electron. Syst.* **2014**, *50*, 1390–1402. [CrossRef]
20. Li, M.K.; Song, D.N. Fault diagnosis method of converter circuit base on wavelet analyze and RF algorithm. *Electr. Eng.* **2016**, *17*, 36–40.
21. Bian, L.; Bian, C.Y. Review on intelligence fault diagnosis in power networks. *Power Syst. Prot. Control* **2014**, *42*, 146–153.
22. Riera-Guasp, M.; Antonino-Daviu, J.A.; Capolino, G.-A. Advances in electrical machine, power electronic, and drive condition monitoring and fault detection: State of the art. *IEEE Trans. Ind. Electron.* **2014**, *62*, 1746–1759. [CrossRef]
23. Shen, G.; Zhou, L.W.; Du, X.; Yang, X.; Xu, M.W. Characteristics analysis of IGBT module with bond wire lift-off based on wavelet singular entropy theory. *Trans. China Electrotech. Soc.* **2013**, *28*, 165–171.
24. Amaral, A.M.; Cardoso, A.M.J. Simple experimental techniques to characterize capacitors in a wide range of frequencies and temperatures. *IEEE Trans. Instrum. Meas.* **2010**, *59*, 1258–1267. [CrossRef]

Disclaimer/Publisher's Note: The statements, opinions and data contained in all publications are solely those of the individual author(s) and contributor(s) and not of MDPI and/or the editor(s). MDPI and/or the editor(s) disclaim responsibility for any injury to people or property resulting from any ideas, methods, instructions or products referred to in the content.

energies

MDPI

Article

Identification and Correction of Abnormal, Incomplete Power Load Data in Electricity Spot Market Databases

Jingjiao Li, Yifan Lv *, Zhou Zhou, Zhiwen Du, Qiang Wei and Ke Xu

School of Electric Power Engineering, Nanjing Institute of Technology, Nanjing 211167, China; jjli@njit.edu.cn (J.L.); y00450220542@njit.edu.cn (Z.Z.)
* Correspondence: x00206191126@njit.edu.cn

Abstract: The development of electricity spot markets necessitates more refined and accurate load forecasting capabilities to enable precise dispatch control and the creation of new trading products. Accurate load forecasting relies on high-quality historical load data, with complete load data serving as the cornerstone for both forecasting and transactions in electricity spot markets. However, historical load data at the distribution network or user level often suffers from anomalies and missing values. Data-driven methods have been widely adopted for anomaly detection due to their independence from prior expert knowledge and precise physical models. Nevertheless, single network architectures struggle to adapt to the diverse load characteristics of distribution networks or users, hindering the effective capture of anomaly patterns. This paper proposes a PLS-VAE-BiLSTM-based method for anomaly identification and correction in load data by combining the strengths of Variational Autoencoders (VAE) and Bidirectional Long Short-Term Memory Networks (BiLSTM). This method begins with data preprocessing, including normalization and preliminary missing value imputation based on Partial Least Squares (PLS). Subsequently, a hybrid VAE-BiLSTM model is constructed and trained on a loaded dataset incorporating influencing factors to learn the relationships between different data features. Anomalies are identified and corrected by calculating the deviation between the model's reconstructed values and the actual values. Finally, validation on both public and private datasets demonstrates that the PLS-VAE-BiLSTM model achieves average performance metrics of 98.44% precision, 94% recall rate, and 96.05% F1 score. Compared with VAE-LSTM, PSO-PFCM, and WTRR models, the proposed method exhibits superior overall anomaly detection performance.

Keywords: anomaly identification and correction; bidirectional long short-term memory network; power load data; partial least square; variational auto-encoders

1. Introduction

The widespread adoption of smart meters and cloud computing technologies has enabled modern power systems to collect and store vast amounts of load data. These data hold significant value for grid companies, electricity retailers, and virtual power plant operators alike [1]. High-quality historical load data are foundational for accurate future load curve prediction. Precise load forecasting is critical for informed decision-making across various aspects of power system operations, including scheduling and control, demand response, and electricity spot market transactions. With the advancement of electricity spot markets, more granular and accurate load forecasting is becoming an essential capability for fine-tuning scheduling control and designing innovative trading products. However, historical load data at the distribution network or user level are frequently plagued by outliers and missing values. These data quality issues arise from factors such as erroneous signal inputs, inaccurate measurements, data extraction problems, communication failures, and integration inconsistencies [2,3]. Poor data quality severely hinders the accurate characterization and prediction of more fine-grained electricity consumption patterns (e.g., at the distribution network or individual user level), thus impacting subsequent decision-making and actions. To avoid the "garbage in, garbage out" problem, it is imperative to identify and rectify missing and outlier values in historical load data to improve data quality. This ensures the reliability of the analysis, prediction, and scheduling processes involved in power system operations and trading [4].

Currently, scholars have researched the identification of anomalous data and have proposed corresponding solutions and measures to improve data quality. These methods can be broadly categorized into four types: (1) statistical methods; (2) distance-based methods; (3) density-based methods; and (4) data-driven methods [5,6]. In the analysis of load data, direct identification and correction of raw daily electricity data often lack a thorough analysis of the underlying components and fail to capture the patterns and influencing factors directly affecting electricity consumption. Statistical methods, including Z scores, box plots, and hypothesis testing [7–13], typically assume that the data follow a specific distribution and contain a certain proportion of normally distributed data. Furthermore, their parameter settings tend to be subjective. Distance-based methods operate on the assumption that normal data points are densely distributed within their local regions while outliers are sparsely distributed [14]. This can be effective for anomaly detection; however, the computational speed slows down when the distance formulas become more complex. Among density-based methods, DBSCAN is a representative technique. It can identify noise points without prior knowledge of the number of clusters to form and can discover clusters of any shape, giving it some applicability. However, it struggles to incorporate influencing factors [15]. Data-driven learning methods, which do not require expert prior knowledge or precise physical models, have also been applied to anomaly detection. However, single-detection network structures often struggle to adapt to the diverse characteristics of distribution grids or user loads and cannot effectively capture data anomaly patterns. In comparison, combined models demonstrate greater potential. For example, the model proposed in [16] combines a novel residual Convolutional Neural Network (CNN) with a layered Echo State Network (ESN) to capture both spatial and temporal dependencies in the data. Furthermore, the Variational Autoencoder (VAE)–Long Short-Term Memory (LSTM)-combined models adopted in [17–19] have achieved effective anomaly detection across multiple data types. These methods primarily learn the patterns of load changes from the training dataset to predict load changes in the test dataset, thus realizing anomaly detection through prediction. However, they often overlook the fact that data after the anomaly detection point already exist and contain a significant amount of pattern information. In reference [20], a bidirectional LSTM (BiLSTM) network was successfully utilized to capture the latent patterns and dependencies in power load data. By encapsulating a contextual understanding of the data, it significantly enhances predictive capabilities and delivers outstanding performance, providing valuable insight. In addition, current approaches to missing value imputation during data preprocessing typically rely on traditional methods such as interpolation, curve fitting, or clustering. These approaches often struggle to consider the underlying influencing factors.

Addressing the aforementioned issues, this paper proposes a method for identifying and correcting missing and anomalous values in power spot market load data. This method is based on a Partial Least Squares (PLS)–Variational Autoencoder (VAE)–Bidirectional Long Short-Term Memory (BiLSTM) network. First, missing values are identified and then initially imputed using a PLS-based approach. This approach considers both the underlying load variation patterns and the influencing factors, thereby preventing the omission of extreme impacts. The completed load curves are then normalized to finalize the data preprocessing step. Second, a VAE-BiLSTM deep hybrid network model is designed, leveraging the representation learning and modeling capabilities of the Variational Autoencoder along with the advantages of the bidirectional long short-term memory network in capturing temporal features and learning contextual information. This model effectively maps relationships between different data features. Finally, the effectiveness of the proposed method is validated through outlier detection and correction experiments conducted on both public and non-public datasets.

2. The Types of Anomalous Power Load Data

In power load data, common types of anomalous data can be broadly categorized into two groups: missing values and outliers. Missing values exhibit a single form, characterized by "NaN" when called. In comparison, outliers appear in more complex forms, such as continuous duplicate data, abnormal peaks (or troughs), or trends that deviate entirely from established electricity usage patterns. In addition to their different manifestations, anomalous data often differ in their causes and handling methods.

2.1. Missing Value

Missing data refer to power load records that were not captured or lost over a specific time point or period. Common causes include equipment failures, sensor disconnections, communication interruptions, or human errors. Handling methods for missing data typically include interpolation, imputation, or predicting missing values using statistical models. The imputation of missing values should aim to preserve the temporal continuity and physical consistency of the data as much as possible.

2.2. Outliers

(1) Continuous Duplicate Data

Continuous duplicate data refer to identical or similar load values recorded over the same period, often caused by sensor malfunctions, data acquisition system failures, or communication delays. These anomalies compromise data validity and may bias load forecasting models. They should be addressed through timestamp inspections or data deduplication algorithms;

(2) Abnormal Peaks (or Troughs)

Abnormal peaks or troughs refer to extreme high or low values in load data that far exceed the normal operating range. These anomalies may arise from equipment failures, sudden load increases or decreases, or environmental factors (e.g., climate changes). Although often transient, such peaks or troughs can significantly impact data analysis and system stability. Detection methods include threshold-based approaches, statistical analysis (e.g., standard deviation), or anomaly detection models;

(3) Abnormal Consumption Trends

Abnormal consumption trends refer to load data patterns or periodic fluctuations that deviate significantly from normal loads. These trends often indicate long-term system issues, equipment aging, or changes in load composition. For instance, load curves may exhibit fluctuations inconsistent with regular operating cycles or sustained deviations. Detecting such anomalies typically relies on time series analysis techniques, such as seasonal adjustments, trend analysis, or machine learning-based pattern recognition methods.

3. The Proposed Model

3.1. Partial Least Squares Regression

Partial Least Squares (PLS) is a regression technique suitable for high-dimensional, multi-collinearity data sets. It can not only extract the principal components of the data but also establish a linear regression relationship between the input variables and the response variables.

The partial least squares regression for the single dependent variable is as follows: With a single dependent variable $Y \in R^n$, independent variable $X = |x_1, x_2, \cdots, x_p|$, $x_j \in R^n$. Partial least squares regression extracts the components t_1 and u_1 in X and Y, respectively, and the extraction of the components is required to satisfy for regression analysis:

(1) t_1 and u_1 should capture as much variation as possible from their respective datasets;
(2) The correlation between t_1 and u_1 must be maximized.

After extracting the first components, t_1 and u_1 partial least squares regression is performed to model X and Y, based on t_1. If the regression equations achieve satisfactory accuracy, the algorithm terminates. Otherwise, a second iteration is conducted using the residuals of X and Y explained by t_1. This process repeats until satisfactory accuracy is reached. If a total of m components $t_1, t_2, \ldots, t_m$ are extracted from X, partial least squares regression is performed by regressing Y on these components. This regression is subsequently expressed as a relationship between Y and the original variables $x_1, x_2, \ldots, x_p$, completing the modeling process.

3.2. Variational Autoencoder Model

Variational Autoencoder (VAE) is a generative model for learning the latent representation of data and generating new data similar to the input data [20]. The objective of VAE is to maximize the similarity between the reconstructed data L' and the input data L while ensuring that the latent variable z follows a predefined prior distribution, typically a standard normal distribution. It consists of an encoder and a decoder. The encoder maps the input data x to the parameters of a distribution in the latent space, e.g., the mean and standard deviation. The latent variable z is then sampled from this distribution. The decoder reconstructs the data $\hat{Y}$ from the sampled latent variable z.

$$z \sim Enc(x) = \mathrm{q}_{\varphi}(z|\mathrm{L}) \tag{1}$$

where φ is the distribution function of the encoder; L is the latent feature, and x is the input function.

The decoding process is the recovery of the hidden variable space data and the decoding, as shown in Equation (2).

$$L' \sim Dec(z) = p_{\theta}(\mathrm{L}|z) \tag{2}$$

where θ is the distribution function of the decoder; L' is the reconstructed data.

Because the potential features z cannot be directly observed in the above process's distribution, through the encoding process $\mathrm{q}_{\varphi}(z|\mathrm{L})$, the posterior distribution of $p_{\theta}(\mathrm{L}|z)$ is replaced.

To approximate the two, the KL scatter degree (the difference between the logarithms of the two distributions) is measured, and the parameter φ, θ are passed to minimize the KL scatter degree, as shown in Equation (3).

$$\begin{aligned} \varphi, \theta &= \mathrm{argmin} D_{KL}\left(\mathrm{q}_{\varphi}(z|\mathrm{L}) \middle|\middle| p_{\theta}(\mathrm{L}|z)\right) \\ &= E_{\mathrm{q}_{\varphi}(z|L)}\left[\log \mathrm{q}_{\varphi}(z|\mathrm{L}) - \log p_{\theta}(\mathrm{L}|z)\right] + \log p_{\theta}(\mathrm{L}) \end{aligned} \tag{3}$$

3.3. Bidirectional Long Short-Term Memory Networks

Unlike load forecasting, the load data before and after the data points to be examined in the anomaly data identification and correction process have already occurred. In this paper, we introduce a bi-directional structure that has advantages in learning the information of front and backward text to improve the LSTM model in order to improve the ability to learn the features of the complete load data. The Bi-LSTM uses two LSTM layers; one reads the window data from the forward direction, and the other one reads the window data from the backward direction and also utilizes the load information before and after the occurrence of each load data, which makes the ability of LSTM model to learn the data features improved. The following Figures 1 and 2 respectively show are the LSTM model structure as well as the Bi-LSTM model structure:

$$f_t = \sigma\left(W_f * [h_{t-1}, x_t] + b_f\right) \tag{4}$$

$$i_t = \sigma(W_i * [h_{t-1}, x_t] + b_i) \tag{5}$$

$$\hat{C}_t = tanh(W_c * [h_{t-1}, x_t] + b_c) \tag{6}$$

$$C_t = f_t * C_{t-1} + i_t * \hat{C}_t \tag{7}$$

$$o_t = \sigma(W_o * [h_{t-1}, x_t] + b_o) \tag{8}$$

$$h_t = o_t * tanh(C_t) \tag{9}$$

where f_t is the output of the forgetting gate; i_t is the control signal of the input gate for controlling the output of the current cell state $\hat{C}_t$; C_t is the current cell state; o_t is the control signal of the output gate, and h_t is the output value of the output gate.

$$y_t = w_{t1} * h_{t1} + w_{t2} * h_{t2} \tag{10}$$

where y_t is the output of the Bi-LSTM in the current window; h_{t1} is the output of the forward LSTM in the current window, and h_{t2} is the output of the backward LSTM's output in the current window.

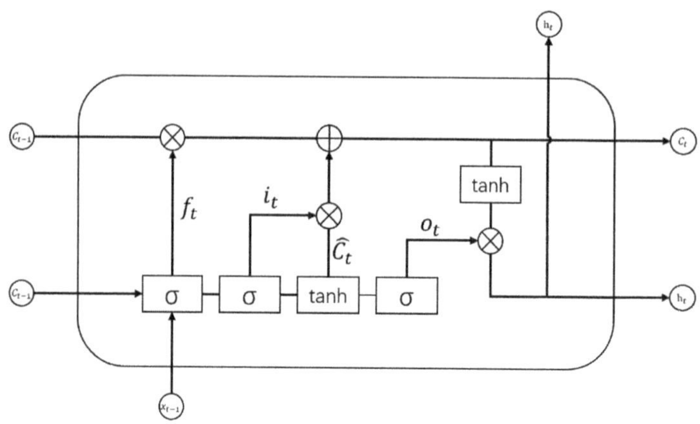

Figure 1. The structure of LSTM network.

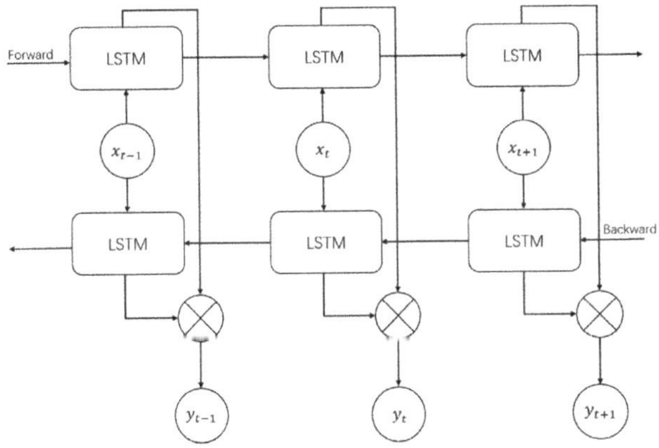

Figure 2. The structure of Bi-LSTM network.

3.4. *VAE-BiLSTM*

In this paper, Bidirectional LSTM (Bidirectional LSTM) is embedded into the encoder and decoder parts of VAE (Variational Autoencoder). Bidirectional LSTM allows the model to learn information in both directions of the time step, which is particularly useful for the processing of time series data.

In the encoder part, each bidirectional LSTM layer can be seen as consisting of two unidirectional LSTM layers, one forward (from t = 1 to t = T) and one backward (from t = T to t = 1).

$$\begin{cases} \overrightarrow{h_t^{(1)}} = \mathrm{LSTM}_1^{\rightarrow}(x_t, \overrightarrow{h_{t-1}^{(1)}}) \\ \overline{h_t^{(1)}} = \mathrm{LSTM}_1^{\leftarrow}(x_t, h_{t+1}^{(1)}) \end{cases} \tag{11}$$

Output:

$$h_t^{(1)} = [\overrightarrow{h_t^{(1)}}; \overrightarrow{h_t^{(1)}}] \tag{12}$$

For the second bidirectional LSTM layer, the input is the output of the first layer:

$$\begin{cases} \overrightarrow{h_t^{(2)}} = \mathrm{LSTM}_2^{\rightarrow}(h_t^{(1)}, \overrightarrow{h_{t-1}^{(2)}}) \\ \overline{h_t^{(2)}} = \mathrm{LSTM}_2^{\leftarrow}(h_t^{(1)}, \overleftarrow{h_{t+1}^{(2)}}) \end{cases} \tag{13}$$

Output:

$$h_t^{(2)} = [\overrightarrow{h_t^{(2)}}; \overleftarrow{h_t^{(2)}}] \tag{14}$$

The decoder is the same. Therefore, the overall structure of BiLSTM embedded in VAE is that the encoder extracts features from the input sequence L through the bidirectional LSTM layer and compresses it into the potential space z, while the decoder decodes the potential space z into the original sequence through the bidirectional LSTM layer.

3.5. *Flowchart*

Figure 3 shows the overall process of identifying and correcting incomplete and abnormal power load data based on the VAE-BiLSTM model. The details are as follows: First, load and influencing factor data are loaded and preprocessed. The preprocessing includes preliminary filling of missing values based on PLS and data normalization. Secondly, based on the representation learning modeling ability of the VAE network and the ability of the bidirectional LSTM to capture long-term and short-term dependencies between contextual data, the bidirectional LSTM is used to replace the BP neural network layer of the traditional VAE, and a VAE-BiLSTM hybrid model for abnormal power load data detection is designed. The model is trained to simulate potential power consumption patterns. The test data $L_{test}(t)$ with anomalies are input into the trained VAE-BiLSTM model, and the model estimates the embedding sequence $Z(t)$ of the test data $L_{test}(t)$ through the VAE encoder; then, the embedding sequence is input into the Bi-LSTM model for self-supervised learning and outputs the reconstructed embedding sequence; finally, the embedding sequence output by the Bi-LSTM is reconstructed through the VAE decoder of the model to obtain the reconstructed result $L'_{test}(t)$ value. The deviation value calculated by input $L_{test}(t)$ and reconstructed output $L'_{test}(t)$ is compared with the threshold to determine whether the load data that have occurred are abnormal. The abnormality is marked and replaced with the reconstructed value.

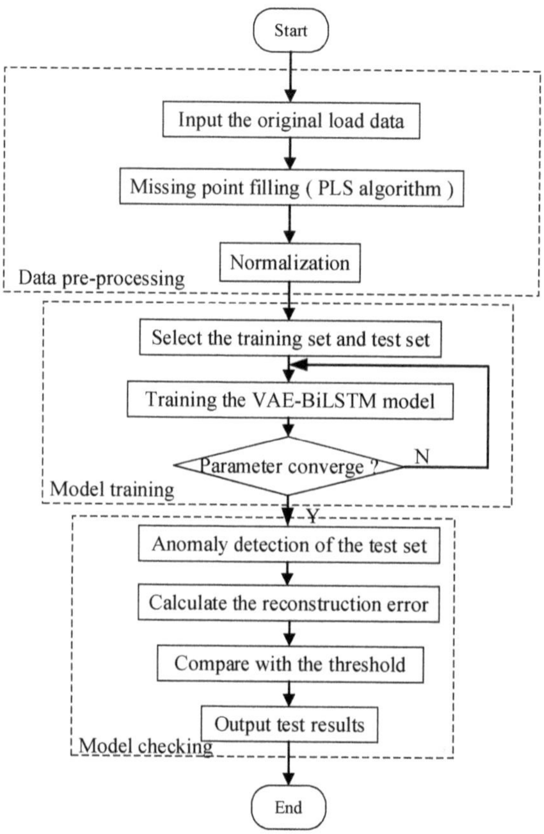

Figure 3. Flowchart of the proposed model.

4. Performance Evaluation Index

In order to evaluate the detection and prediction accuracy of the model for the user's daily load, this study uses precision, recall, and F1 scores as indicators. The calculation expressions are as follows:

$$\text{Precision} = \frac{\text{TP}}{\text{TP} + \text{FP}} \tag{15}$$

$$\text{Recall} = \frac{\text{TP}}{\text{TP} + \text{FN}} \tag{16}$$

$$F1 = 2 \times \frac{\text{Precision} \times \text{Recall}}{\text{Precision} + \text{Recall}} \tag{17}$$

Among them, TP (True Positive) represents the number of samples that correctly detect anomalies, that is, the number of samples with an anomaly detection label of 1 and a real label of 1; TN (True Negative) represents the number of samples that are correctly detected as normal, that is, the number of samples with an anomaly detection label of 0 and a real label of 0. FP (False Positive) represents the number of misdetected samples, that is, the number of normal samples misjudged as abnormal samples; FN (False Negative) represents the number of missed samples, that is, the number of abnormal samples that cannot be detected; precision is an indicator for evaluating the accuracy of the algorithm model in detecting positive samples; the larger, the better; recall is an indicator for evaluating the ability of the algorithm model to detect the full degree of positive samples, the bigger the better; precision is mutually restricted with the recall. The comprehensive indicator F1 score is introduced for evaluation; the bigger, the better.

5. Case Analysis

5.1. Test Platform and Data Sources

The PLS-VAE-Bi LSTM model proposed in this paper is built using deep learning Pytorch running on PyCharm 2024.3.1.1. The CPU is Intel Core i7-9750H CPU 2.60 GHz, 16 GB memory, and the graphics card is NVIDIA GTX1650, 4 GB video memory.

This paper uses public and private power load datasets to test the algorithm, including main grid load, distribution grid load [21], manufacturing user's load, and a residential user's load [22]. The data set covers not only the load data and meteorological factors of different grid levels but also the load data of power users in different industries. The model is trained using historical two-month load data, and the trained model is used to detect abnormal load data in the next 10 days (once every 30 min, 48 data points per day).

5.2. Network Training and Result Analysis

The trend of the loss function of the PLS-VAE-Bi LSTM model is shown in Figure 4.

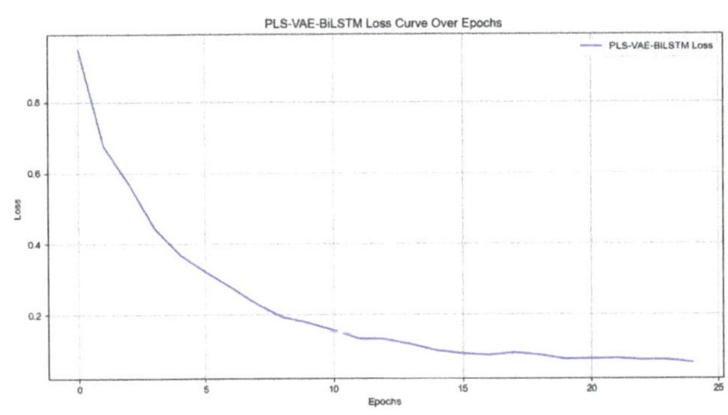

Figure 4. Trend chart of load data loss function.

In the initial stage of training, the loss of PLS-VAE-BiLSTM is high, and the model has not learned the distribution of data well. In the previous epochs, the loss decreases rapidly, indicating that the model is learning quickly and adjusting its parameters to better fit the data. In the next training process, the loss tends to be stable, indicating that the model has basically converged, and further training has little effect on the loss. The rapid decrease in VAE loss indicates that the PLS-VAE-BiLSTM model quickly learns the distribution of data at the initial stage and then tends to be stable, indicating that the model has basically converged.

As shown in Figure 5, the analysis of the abnormal load data identification results across the manufacturing, residential, distribution grid, and main grid reveals that the F1 index initially increases and then sharply decreases within the threshold range of 80% to 100%. Specifically, the F1 index for the manufacturing user increases gradually with the threshold, peaking at 0.9474 at 98% before rapidly declining. In contrast, the residential user reaches its highest value of 0.9677 at 96.5%, indicating optimal identification performance. The distribution grid shows consistent improvement, with the F1 index steadily increasing and reaching a peak of 0.9796 at 97.5%, the highest among the four sectors. Meanwhile, the main grid F1 index peaks at 0.9474 at 96%, after which a further threshold increase leads to a significant decline. These results demonstrate that the PLS-VAE-BiLSTM-based abnormal load data identification method proposed in this study performs well across various thresholds, with a significant improvement in the F1 index near the optimal threshold. The model achieves efficient and stable identification across all industries, validating its applicability and strong generalization in multi-industry contexts.

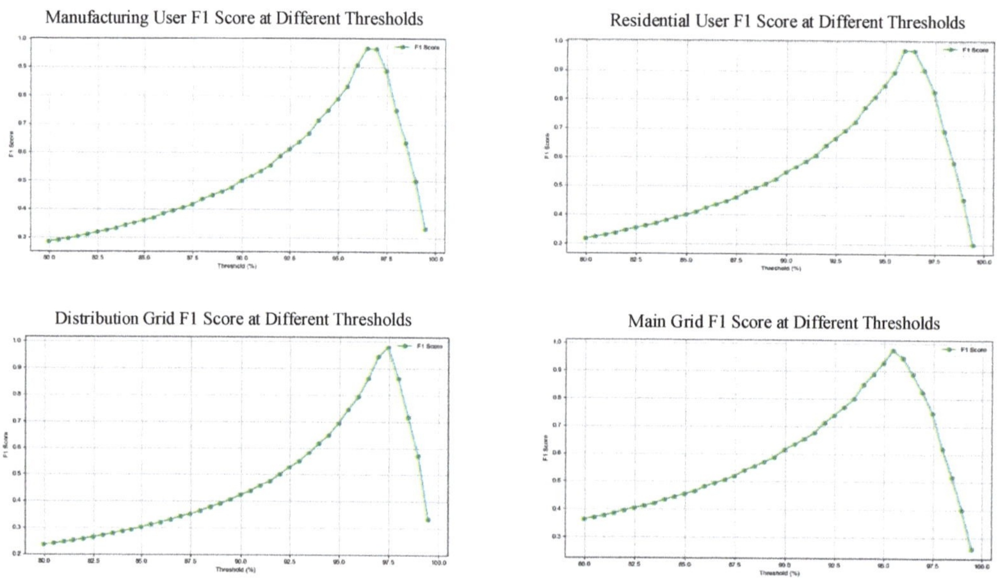

Figure 5. F1 score of abnormal load identification under different thresholds.

5.3. Comparison Experiments

Figure 6 compares the performance of the PLS-VAE-BiLSTM model proposed in this study with the VAE-LSTM, PSO-PFCM [23], and WTRR [24] models in detecting abnormal load data in the manufacturing user. The results show that the accuracy of the PLS-VAE-BiLSTM model is 100.00%, significantly outperforming VAE-LSTM (95%), PSO-PFCM (92%), and WTRR (90%). The recall rate is 90.00%, balancing the avoidance of excessive false detections with a high recall rate. The F1 index reaches 94.74%, significantly outperforming other models (VAE-LSTM: 89%, PSO-PFCM: 87.20%, WTRR: 84.21%), demonstrating a strong balance between accurate detection and comprehensive coverage of abnormal data. Manufacturing user load data exhibit strong periodicity and complex fluctuations, with abnormal loads frequently occurring during peak production periods. Traditional methods, such as PSO-PFCM and WTRR, struggle to handle nonlinear load fluctuations, leading to a lower F1 index.

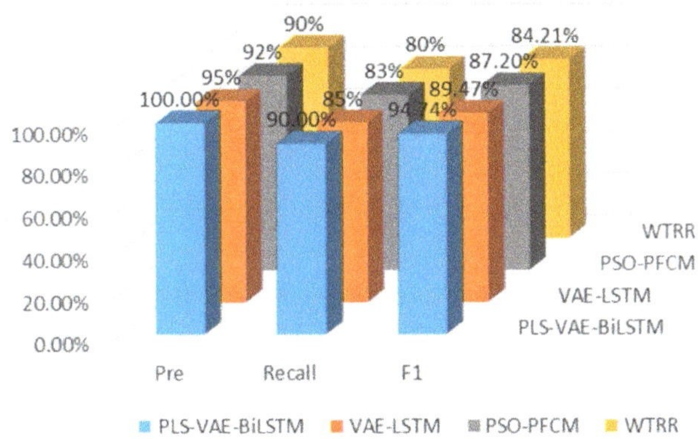

Figure 6. Comparison of abnormal load identification results in manufacturing user load dataset.

Figure 7 compares the performance of the PLS-VAE-BiLSTM model proposed in this study with the VAE-LSTM, PSO-PFCM, and WTRR models in detecting abnormal loads in the residential user. The results show that the PLS-VAE-BiLSTM model exhibits clear advantages in accuracy, recall, and F1 scores. Specifically, the accuracy of the PLS-VAE-BiLSTM model is 93.75%, outperforming VAE-LSTM (90%), PSO-PFCM (85%), and WTRR (80%). The recall rate is 100.00%, significantly higher than the other models, indicating that PLS-VAE-BiLSTM captures abnormal data more comprehensively. In terms of F1 score, PLS-VAE-BiLSTM achieved 96.77%, surpassing VAE-LSTM (92.11%), PSO-PFCM (87.62%), and WTRR (82.61%), demonstrating an effective balance between high-precision detection and a low false-positive rate.

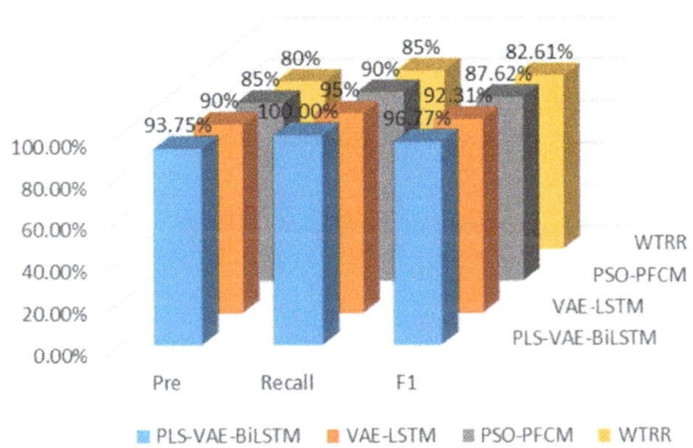

Figure 7. Comparison of abnormal load identification results in residential user load dataset.

Figure 8 illustrates the performance of the PLS-VAE-BiLSTM model proposed in this study, alongside the VAE-LSTM, PSO-PFCM, and WTRR models, in detecting abnormal loads in the distribution grid. The results show that the PLS-VAE-BiLSTM model significantly outperforms the other models in terms of accuracy, recall, and F1 score. Specifically, the accuracy of PLS-VAE-BiLSTM is 100%, surpassing VAE-LSTM (97%), PSO-PFCM (93%), and WTRR (85%). Its recall rate is 96%, significantly higher than the other models, indicating VAE-BiLSTM's superior ability to capture and identify abnormal loads. The F1 score of PLS-VAE-BiLSTM is 97.96%, leading VAE-LSTM (94%), PSO-PFCM (90%), and WTRR (82.35%), demonstrating an effective balance between accurate detection and a low false-positive rate. In contrast, the PSO-PFCM and WTRR models struggled to handle the random fluctuations and sudden changes in load within the complex distribution grid, leading to lower detection accuracy and recall rates compared to PLS-VAE-BiLSTM. PLS-VAE-BiLSTM demonstrates significant improvement in abnormal load detection in the distribution grid, owing to its superior nonlinear feature extraction and time-series dependency capture abilities, showcasing its strong adaptability and advantages in power system load monitoring.

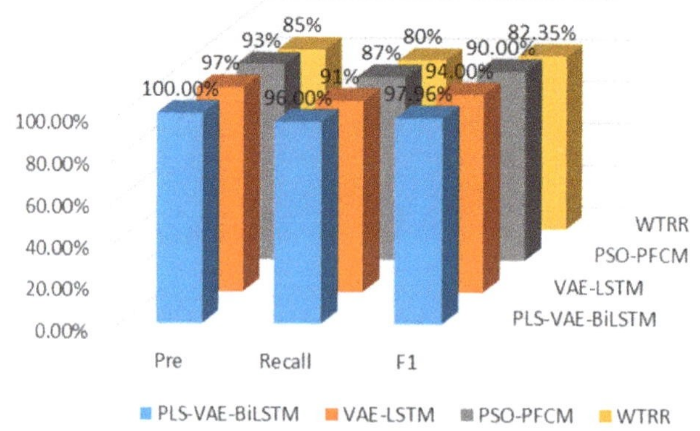

Figure 8. Comparison of abnormal load identification results in distribution grid load dataset.

Figure 9 compares the performance of the PLS-VAE-BiLSTM model proposed in this study with the VAE-LSTM, PSO-PFCM, and WTRR models in detecting main grid load anomalies. The results demonstrate that the PLS-VAE-BiLSTM model significantly outperforms the other models in accuracy, recall, and F1 score. Specifically, PLS-VAE-BiLSTM achieves 100% accuracy, surpassing VAE-LSTM (98%), PSO-PFCM (92%), and WTRR (88%). Its recall rate is 90.00%, significantly higher than the other models, indicating its superior ability to capture abnormal data. The F1 score of PLS-VAE-BiLSTM is 94.74%, outperforming VAE-LSTM (91.84%), PSO-PFCM (85.71%), and WTRR (81.63%), demonstrating an excellent balance between high accuracy and low false-positive rate. Unlike anomalies in the distribution grid and residential industries, main gird load anomalies typically manifest as widespread trend deviations, significantly impacting power system stability and exhibiting strong global and long-term characteristics. In contrast, although PSO-PFCM and WTRR can handle some load fluctuations, they lack the nonlinear modeling and time-series processing capabilities required to address the complex global anomalies of the main gird load, resulting in inferior performance compared to PLS-VAE-BiLSTM in accuracy and recall.

As illustrated in Figure 10, the PLS-VAE-BiLSTM model demonstrates superior performance in the anomaly detection task, achieving a precision of 98.44%, a recall of 94%, and an F1 score of 96.05%. This highlights its excellent ability to accurately capture anomalies while maintaining high precision. The VAE-LSTM model follows closely, with a precision of 94.88%, a recall of 89%, and an F1 score of 91.91%, reflecting strong overall performance. In comparison, the PSO-PFCM model shows moderate performance, achieving a precision of 90.5%, a recall of 85%, and an F1 score of 87.63%, which are slightly lower than those of the VAE-based models. The WTRR model performs the worst, with a precision of 85.75%, a recall of 80%, and an F1 score of 82.7%, indicating limited anomaly capture capability and a higher false alarm rate. Overall, the PLS-VAE-BiLSTM model outperforms all other compared models, showcasing its clear advantage in anomaly detection tasks.

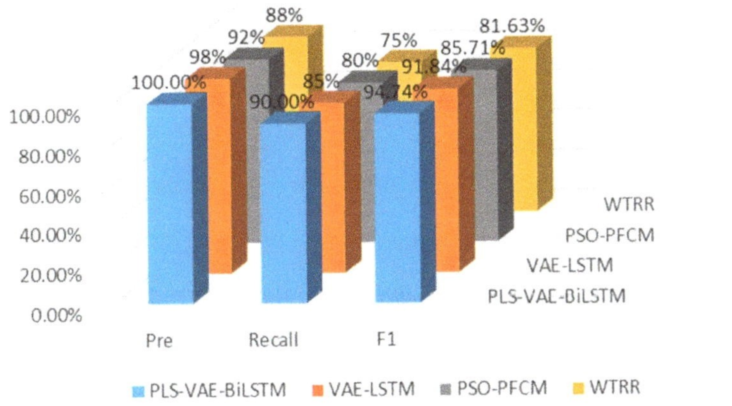

Figure 9. Comparison of abnormal load identification results in main grid load dataset.

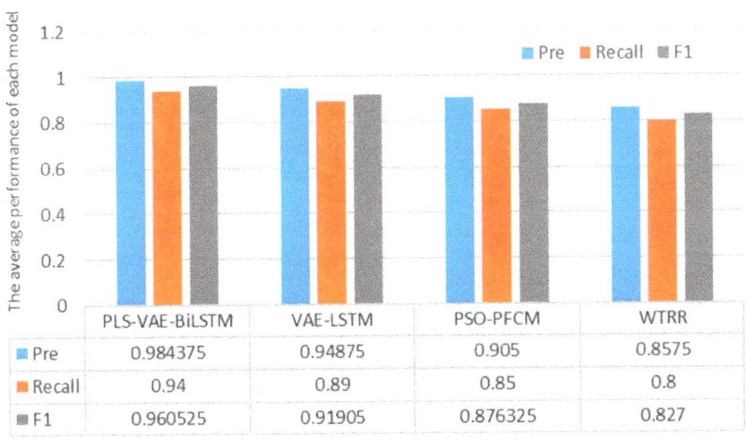

Figure 10. The average performance of each model.

6. Discussion

Currently, the volume of fine-grained power load data is rapidly increasing, and the value of this data is contingent upon its quality meeting application requirements. To enhance anomaly detection in fine-grained load data from distribution networks or users and address the limitations of single-network models in capturing data patterns or accounting for sequential data relationships, this paper has proposed a method for identifying and correcting missing and anomalous load data based on a PLS-VAE-BiLSTM hybrid model. This method introduces the PLS approach for missing value preprocessing and combines it with the representation learning capabilities of the VAE and the contextual modeling advantages of the BiLSTM network. First, data preprocessing is performed, including PLS-based missing value imputation and data normalization. Second, a VAE-BiLSTM hybrid model is constructed and trained using a loaded dataset incorporating influencing factors. This training process allows the model to learn the relationships between different data features. Anomalous data are then identified and corrected by calculating the deviations between the model's reconstructed values and the actual values. Finally, the proposed method is validated on four real-world datasets. The experimental results demonstrate that the VAE-BiLSTM model achieves average performance metrics of 98.44% for precision, 94% for recall, and 96.05% for F1 score. In comparison to VAE-LSTM, PSO-PFCM, and WTRR models, the proposed model exhibits superior overall anomaly detection performance. Specifically, the VAE-BiLSTM model achieves 3.75%, 8.77%, and 14.79% higher precision than the VAE-LSTM, PSO-PFCM, and WTRR models, respectively. It also demonstrates 5.62%, 10.59%, and 17.5% higher recall than those models. Furthermore, the F1 score is 4.51%, 9.6%, and 16.15% higher compared to the VAE-LSTM, PSO-PFCM, and WTRR models, respectively. For each method, the precision and recall results across different datasets show a characteristic trade-off relationship. Overall, the experiments demonstrate that the PLS-VAE-BiLSTM hybrid model can effectively capture the interdependencies within load data and between load data and their influencing factors, exhibiting strong overall detection performance.

In the current work, the single-dependent-variable PLS model can establish a maximized linear regression model between key influencing factors and the load. This is computationally efficient but is limited by its linear form and inability to decouple the correlations between influencing factors, which requires further improvements in future research. In the VAE-BiLSTM hybrid model, the threshold settings for anomaly detection with VAE, which are dataset-dependent and learned through training, need to be considered regarding the model training time and resource consumption when deploying the model on a platform. Future work will focus on optimizing the model performance, making it lightweight, and reducing the detection time.

7. Conclusions

Based on an analysis of the issues and limitations of existing methods, this paper proposes a power load anomaly detection and correction method based on the PLS-VAE-BiLSTM model. The conclusions and recommendations are summarized as follows:

1. Data Preprocessing:

The Partial Least Squares (PLS) method was used to establish a maximized linear regression model between historical load and influencing factors. This approach incorporates influencing factors to impute missing values, forming a complete time-series load curve that undergoes normalization for further analysis;

2. Anomaly Identification and Correction:

The VAE-BiLSTM model is trained to simulate historical data trends for anomaly detection. In this model, BiLSTM replaces the BP neural network in the VAE framework for encoding and decoding, effectively integrating influencing factors such as calendar and weather data. The model learns the features of load data through training, reconstructs the data to capture load variation trends, and identifies anomalies by comparing reconstructed data with actual data for deviation correction. Compared to LSTM models used in previous predictive studies, the BiLSTM model demonstrates superior performance in anomaly cleaning due to its ability to better leverage contextual information from both preceding and succeeding data;

3. Consideration of Influencing Factors and Future Improvements:

The proposed model considers major factors affecting power loads, such as calendar and weather, during both data preprocessing and cleaning stages. However, there is room for further improvement in algorithm design and computational performance. Additionally, the exploration of more influencing factors holds potential for enhancing the model's effectiveness.

Author Contributions: Conceptualization, J.L.; software, Y.L.; validation, Z.Z.; formal analysis, Z.D. and Q.W.; writing—original draft, Y.L.; writing—review and editing, J.L.; supervision, K.X. All authors have read and agreed to the published version of the manuscript.

Funding: This research was funded by the Talent Introduction Scientific Research Foundation of Nanjing Institute of Technology: Research on the Mechanism of Demand Side Response Resources Participating in the Electricity Market (No. YKJ202010).

Data Availability Statement: The original contributions presented in this study are included in this article. Further inquiries can be directed to the corresponding author.

Conflicts of Interest: The authors declare no conflicts of interest.

References

1. Gope, P.; Sikdar, B. Lightweight and Privacy-Friendly Spatial Data Aggregation for Secure Power Supply and Demand Management in Smart Grids. *IEEE Trans. Inf. Forensics Secur.* **2019**, *14*, 1554–1566. [CrossRef]
2. Li, Y.; Yang, R.; Guo, P. Spark-based Parallel OS-ELM Algorithm Application for Short-term Load Forecasting for Massive User Data. *Electr. Power Compon. Syst.* **2020**, *48*, 603–614. [CrossRef]
3. Sohei, I.; Takayuki, N.; Yusuke, K.; Masahiro, M.; Koji, Y. Distillation-Based Semi-Supervised Federated Learning for Communication-Efficient Collaborative Training with Non-IID Private Data. *IEEE Trans. Mob. Comput.* **2021**, *22*, 191–205.
4. Thomson, V.E. *Garbage in, Garbage Out: Solving the Problems with Long-Distance Trash Transport*; University of Virginia Press: Charlottesville, VA, USA, 2009; pp. 1–173.
5. Barnett, V.; Lewis, T.; Abeles, F. Outliers in statistical data. *Phys. Today* **1979**, *32*, 73–74. [CrossRef]
6. Zhuo, L.; Zhao, H.; Zhan, S. Overview of Anomaly Detection Methods and Their Applications. *Comput. Appl. Res.* **2020**, *37*, 9–15.
7. Mendenhall, W.M.; Sincich, T.L. *Statistics for Engineering and the Sciences*; CRC Press: Boca Raton, FL, USA, 2016.
8. Bury, K. *Statistical Distributions in Engineering*; Cambridge University Press: Cambridge, UK, 1999.
9. Tong, S.; Wen, F.; Chen, L.A. Two-dimension Wavelet Threshold De-noising Method for Electric Load Data Pre-processing. *Autom. Electr. Power Syst.* **2012**, *36*, 101–105.
10. Wang, H.; Bah, M.J.; Hammad, M. Progress in Outlier Detection Techniques: A Survey. *IEEE Access* **2019**, *7*, 1. [CrossRef]
11. Li, C. Preprocessing Methods and Pipelines of Data Mining: An Overview. *arXiv* **2019**, arXiv:1906.08510.
12. Patel, V.; Kapoor, A.; Sharma, A.; Chakrabarti, S. Taxonomy of outlier detection methods for power system measurements. *Energy Convers. Econ.* **2023**, *4*, 73–88. [CrossRef]
13. Simmons, J.P.; Nelson, L.D.; Simonsohn, U. False-positive psychology: Undisclosed flexibility in data collection and analysis allows presenting anything as significant. *Psychol. Sci.* **2011**, *22*, 1359–1366. [CrossRef] [PubMed]
14. Knox, E.M.; Ng, R.T. Algorithms for mining distance based outliers in large datasets. In Proceedings of the International Conference on Very Large Data Bases, New York, NY, USA, 24–27 August 1998; pp. 392–403.
15. Schubert, E.; Sander, J.; Ester, M.; Kriegel, H.P.; Xu, X. DBSCAN revisited, revisited: Why and how you should (still) use DBSCAN. *ACM Trans. Database Syst. (TODS)* **2017**, *42*, 1–21. [CrossRef]
16. Alanazi, M.D.; Saeed, A.; Islam, M.; Habib, S.; Sherazi, H.I.; Khan, S.; Shees, M.M. Enhancing Short-Term Electrical Load Forecasting for Sustainable Energy Management in Low-Carbon Buildings. *Sustainability* **2023**, *15*, 16885. [CrossRef]
17. Lin, S.; Clark, R.; Birke, R.; Schönborn, S.; Trigoni, N.; Roberts, S. Anomaly Detection for Time Series Using VAE-LSTM Hybrid Model. In Proceedings of the ICASSP 2020—2020 IEEE International Conference on Acoustics, Speech and Signal Processing (ICASSP), Barcelona, Spain, 4–8 May 2020; pp. 4322–4326. [CrossRef]
18. Wang, C.; Zhang, A.; Yang, L.; Zhang, B.; Li, S. Anomaly detection of UAV flight data based on VAE-LSTM modeling. *Electron. Meas. Technol.* **2024**, *47*, 187–196. [CrossRef]
19. Jing, Z.; Chai, L.; Hu, S. Research on abnormal load detection method for distribution network based on improved LSTM-VAE. *Electr. Meas. Instrum.* **2024**, *61*, 71–76. [CrossRef]
20. Pavlatos, C.; Makris, E.; Fotis, G.; Vita, V.; Mladenov, V. Enhancing Electrical Load Prediction Using a Bidirectional LSTM Neural Network. *Electronics* **2023**, *12*, 4652. [CrossRef]
21. Available online: https://archive.ics.uci.edu/dataset/849/power+consumption+of+tetouan+city (accessed on 1 June 2024).
22. Available online: https://archive.ics.uci.edu/dataset/235/individual+household+electric+power+consumption (accessed on 23 March 2022).
23. Li, Q. Power big data anomaly detection method based on an improved PSO-PFCM clustering algorithm. *Power Syst. Prot. Control.* **2021**, *49*, 161–166. [CrossRef]
24. Karkhaneh, M.; Ozgoli, S. Anomalous load profile detection in power systems using wavelet transform and robust regression. *Adv. Eng. Inform.* **2022**, *53*, 101639. [CrossRef]

Disclaimer/Publisher's Note: The statements, opinions and data contained in all publications are solely those of the individual author(s) and contributor(s) and not of MDPI and/or the editor(s). MDPI and/or the editor(s) disclaim responsibility for any injury to people or property resulting from any ideas, methods, instructions or products referred to in the content.

energies

MDPI

Article

Analysis and Control Parameters Optimization of Wind Turbines Participating in Power System Primary Frequency Regulation with the Consideration of Secondary Frequency Drop

Ketian Liu *, Zhengxi Chen, Xiang Li and Yi Gao

School of Electric Power Engineering, Nanjing Institute of Technology, Nanjing 211167, China; y00450230710@njit.edu.cn (Z.C.); y00450200529@njit.edu.cn (X.L.); y00450220619@njit.edu.cn (Y.G.)
* Correspondence: ktliu@njit.edu.cn

Abstract: With the increasing integration of wind energy into power systems, maintaining frequency stability has become a significant challenge. To address the issue of secondary frequency drop caused by wind turbines exiting the primary frequency regulation of power systems, this paper presents a control parameters optimization method of wind turbines participating in power system primary frequency regulation. Initially, with the assumption of constant wind speed and linearization of the wind power coefficient, the relationship between the mechanical power and rotor speed of the wind turbines is established. Subsequently, the primary frequency regulation component of wind turbines is integrated into the classical system frequency response (SFR) model, accounting for the effects of exiting time and rotor speed variations. Following this, the dynamic frequency of the power system is computed with the modified SFR model, and the time domain expressions for both primary and secondary frequency drops are derived. Furthermore, an optimization model for the control parameters of wind turbines participating in primary frequency regulation is developed, aiming to minimize the values both of primary and secondary frequency drops. Finally, a case study is constructed to validate the efficacy of the proposed method. The results demonstrate that the optimization method introduced in this paper significantly enhances the dynamic characteristics of the system frequency.

Keywords: secondary frequency drop; wind turbines; primary frequency regulation; control parameter optimization; system frequency response model

1. Introduction

Driven by the global energy transition, renewable energy sources, particularly wind and photovoltaic power, are progressively displacing conventional power generation on an unprecedented scale worldwide. As of the end of 2024, China's installed capacity of new energy power generation, predominantly consisting of wind and photovoltaic power, had reached 1.45 billion kW, marking a historic milestone by surpassing thermal power capacity for the first time, while demonstrating accelerated growth in the renewable energy sector. The integration of large-scale wind power, photovoltaic power, and other renewable energy sources into the power system via power electronic converters has led to the gradual displacement of conventional synchronous generator units. This transition has resulted in a reduction in overall power system inertia, a decline in primary frequency regulation capability, and a deterioration of dynamic frequency characteristics. As a consequence, the power system's resilience to power disturbances has been significantly compromised, heightening the risk of frequency instability [1].

Grid-connected wind power and photovoltaic power should have active primary frequency regulation ability with the increasing penetration rate of new energy [2]. Wind turbines participate in the primary frequency regulation of a power system by utilizing the rapid power modulation capability of electronic converters. The control strategies primarily include virtual inertia control and droop control [3]. Virtual inertia control emulates the inertial response of conventional synchronous generator units, while droop control replicates the governor characteristics of these traditional units [4,5]. Concurrently, both pitch angle control and de-loaded power tracking control are employed to supply power reserve for wind turbines' primary frequency regulation [6]. Nevertheless, these methods lead to decreased wind energy utilization efficiency, thereby exhibiting economic drawbacks.

The release of more than half of the rotor's kinetic energy by wind turbines would result in improved economic performance and greater potential in frequency regulation [7]. However, most existing studies have failed to account for the dynamic changes in wind turbine rotor speed and kinetic energy during primary frequency regulation. The release of kinetic energy from wind turbines results in a reduction in rotor speed and mechanical power [8], which imposes temporal constraints on wind turbines' engagement in primary frequency regulation control. When wind turbines exit primary frequency regulation control due to the rotor speed decline, a new active power deficit will be caused in the power system, resulting in the secondary frequency drop and affecting the frequency stability of power system [9]. The previously mentioned problem can be mitigated by implementing optimal control strategies for parameters adjustment and power regulation in wind turbines engaged in primary frequency regulation. These approaches are designed to mitigate the power deficit resulting from wind turbines' disengagement from primary frequency regulation. The specific measures comprise (1) the enhancement of the power–rotor speed characteristics during primary frequency regulation [10–12]; (2) optimization of primary frequency regulation control parameters based on disturbance power magnitude to prevent excessive release of rotor kinetic energy [13–15]; and (3) implementation of advanced control strategies, including (i) nonlinear droop control with adaptive parameter adjustment according to rotor speed variations [16], (ii) utilization of energy state indices for rotor kinetic energy quantification and corresponding parameter optimization [17], and (iii) coordinated operation with energy storage systems to maintain continuous frequency regulation capability [18–20].

In contrast to previous studies, this paper focuses on the optimization of primary frequency regulation control parameters for wind turbines, with particular attention to the secondary frequency drop phenomenon. Unlike the previous studies, we integrate both wind turbine inertia control and droop control into the system frequency response (SFR) model. This dual-control integration enables a more comprehensive analysis of the dynamic relationship between wind turbine output power, rotor speed, and system frequency characteristics. The modified SFR model provides a more precise assessment of rotor kinetic energy's role in frequency regulation processes. Furthermore, we introduce a novel consideration of wind turbine exiting time into the SFR model, allowing for systematic simulation of its impact on secondary frequency drop characteristics. These methodological advancements in the modified SFR model offer unprecedented insights into the kinetic energy release dynamics and their subsequent effects on secondary frequency drop that have been insufficiently addressed in existing research paradigms.

The primary objective of this paper is to develop and optimize the primary frequency regulation control parameters for wind turbines, with a specific focus on minimizing both primary and secondary frequency drops. An optimization approach is proposed to adjust the control parameters of wind turbines in such a way that both primary and secondary frequency drop values are minimized, resulting in a more efficiently viable primary frequency regulation process. Thus, the optimization model presented shifts the focus from merely delaying the secondary frequency drop to integrating it into the optimization process of primary frequency regulation. The effectiveness of wind turbine participation in power system frequency regulation can be enhanced.

2. Analysis of the Influence of the Wind Turbine Rotor Speed on Mechanical Power

The wind turbine mechanical power is affected by two key factors: the wind power coefficient and the input wind power. Assuming that there are n_{w} wind turbines participating in the primary frequency regulation, the total mechanical power of the wind turbines can be expressed as follows:

$$P_{\mathrm{m}} = C_{\mathrm{p}} \cdot P_{\mathrm{v}} \cdot n_{\mathrm{w}} \tag{1}$$

where P_{m} is the total mechanical power of wind turbines; C_{p} is the wind power coefficient; and P_{v} is the input wind power of single wind turbine.

The expression of P_{v} is

$$P_{v} = \frac{1}{2}\rho S_{\mathrm{w}} v^{3} \tag{2}$$

where ρ is the air density; S_{w} is the wind turbine blade swept area; and v is the wind speed.

The wind power coefficient is affected by the pitch angle and tip speed ratio [21]. In the absence of pitch angle control, the wind turbine power coefficient C_{p} can be expressed in terms of the tip speed ratio λ as follows:

$$C_{\mathrm{p}}(\lambda) = C_{1}\left[C_{2}\left(\frac{1}{\lambda} - 0.035\right) - C_{3}\right] \cdot \exp\left[-C_{4}\left(\frac{1}{\lambda} - 0.035\right)\right] + C_{5}\lambda \tag{3}$$

where $C_1 \sim C_5$ are coefficients related to the wind turbine power coefficient characteristics; λ is the tip speed ratio.

The wind turbine tip speed ratio can be expressed as follows:

$$\lambda = \frac{R_{\mathrm{w}}\omega_{\mathrm{r}}}{v} \tag{4}$$

where R_{w} is the wind turbine blade radius. ω_{r} is the wind turbine rotor speed.

The variations in wind turbine rotor speed and wind speed lead to a corresponding variation in tip speed ratio. The small signal increment of C_{p} is

$$\Delta C_{\mathrm{p}}(\omega_{r}) = \frac{0.29\pi R_{w}}{V_{w}} \cdot \cos\left(\frac{\pi R \omega_{r0}}{15V_{w}} - \frac{\pi}{5}\right)\Delta\omega_{r} \tag{5}$$

where λ_0 is the initial tip speed ratio.

In the context of the primary frequency regulation of wind turbines, the influence of wind speed change is often overlooked. By substituting Equations (4) and (5) into Equation (1), the mechanical power change in the wind turbines can be determined:

$$\Delta P_{\mathrm{m}} = C_{\mathrm{p}}'(\omega_{r0}) \cdot P_{v} \cdot n_{\mathrm{w}} \cdot \Delta\omega_{\mathrm{r}} = K_{\mathrm{cp}} \cdot \Delta\omega_{\mathrm{r}} \tag{6}$$

where K_{cp} is the ratio between the mechanical power change and rotor speed change in wind turbines. $\Delta\omega_{\mathrm{r}}$ is the rotor speed change.

3. Analysis of Wind Turbine Rotor Motion Status When Participating in Frequency Regulation

Figure 1 illustrates the wind turbine power–rotor speed curves participating in the primary frequency regulation. In the absence of de-loaded power tracking control, the initial operating point of the wind turbine is point A of the intersection between the mechanical power curve and the maximum power point tracking (MPPT) curve. As the power system load increases, the wind turbine generates additional electromagnetic power in accordance with the control strategy. Concurrently, the electromagnetic power exceeds the mechanical power, resulting in a reduction in rotor speed and a shift in the operating point from point A to point B. When the wind turbine quits the primary frequency regulation, the electromagnetic power abruptly declines to the MPPT power corresponding to the current rotor speed, resulting in the wind turbine transitioning from operating point B to point C. At point C, the mechanical power exceeds the electromagnetic power, the wind turbine rotor speed increases, and the operating point gradually ascends to point A along the curve CA.

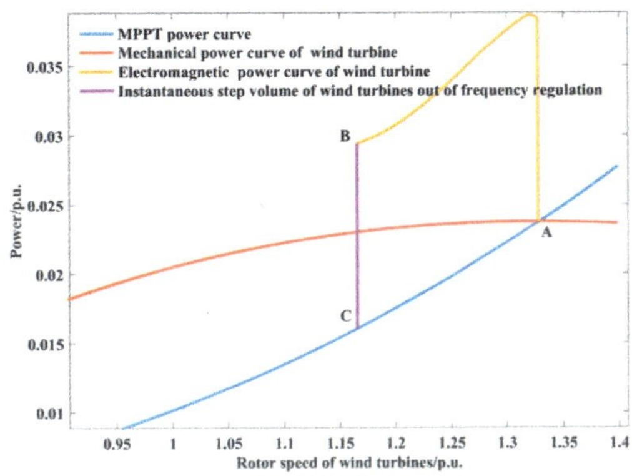

Figure 1. The relationship curves between the power and rotor speed of the wind turbine.

Considering the interaction between system frequency and the wind turbine's active power, as well as the influence of the wind turbine's participation in primary frequency regulation on the wind turbine rotor speed, the system frequency response model, which incorporates the wind turbine's primary frequency regulation link introduced into the SFR model, is illustrated in Figure 2.

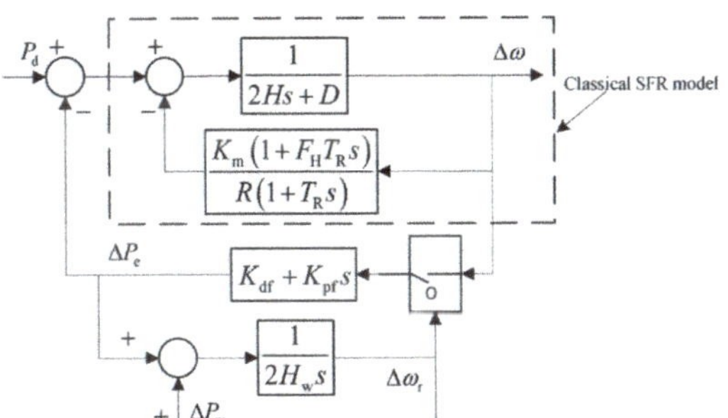

Figure 2. Modified SFR model with primary frequency regulation of wind turbine.

In this figure, H is the inertia constant of the conventional generator unit. D is the damping coefficient. K_m is the mechanical power gain coefficient. R is the governor drop coefficient. F_H is the fraction of total power generated by the high-pressure turbine. T_R is the reheater time constant. P_d is the disturbance power caused by load changing. The value greater than zero indicates a decrease in load, while the value less than zero indicates an increase in load. $\Delta\omega$ is rotor speed variation in the traditional generator rotor. H_w is the inertia time constant of the wind turbine under the equivalent value of a single unit.

K_{df} is the droop control coefficient. K_{pf} is the virtual inertia control coefficient. ΔP_e is the variation in electromagnetic power of the wind turbine. The value less than zero indicates an increase in electromagnetic power, while the value greater than zero indicates a decrease. ΔP_m is the variation in mechanical power of the wind turbine. The switch O is a switch for the wind turbine to participate in a frequency modulation. The switch is closed when the wind turbine is participating in the primary frequency regulation. Conversely, when the wind turbine is no longer participating in the primary frequency regulation, the switch is turned off.

With system per unit, the system frequency variation is equivalent to the rotor speed variation in the traditional generator, and the blade speed is equivalent to the rotor speed of the wind turbine.

Assuming that the disturbance power P_d is a step function and the step amount is P_{step}, the system frequency change during the primary frequency regulation of the wind turbine is obtained from Figure 2:

$$\begin{aligned}\Delta f(s) &= \Delta\omega(s)\\ &= \frac{RP_{step}}{DR+K_m+RK_{df}}\cdot\frac{(1+T_R s)\omega_n^2}{s(s^2+2\zeta\omega_n s+\omega_n^2)}\end{aligned} \quad (7)$$

where

$$\begin{cases}\omega_n^2 = \frac{DR+K_m+RK_{df}}{2HRT_R+RK_{pf}T_R}\\ \zeta = \frac{2HR+(DR+K_m F_H+RK_{df})T_R+RK_{pf}}{2(DR+K_m+RK_{df})}\cdot\omega_n\end{cases} \quad (8)$$

In Figure 2, the change in electromagnetic power output by the wind turbine participating in the primary frequency regulation can be expressed as

$$\begin{aligned}\Delta P_e(s) &= \Delta\omega(s)K_{df} + \Delta\omega(s)K_{pf}s\\ &= \Delta P_{K_{df}}(s) + \Delta P_{K_{pf}}(s)\end{aligned} \quad (9)$$

where ΔP_{Kdf} and ΔP_{Kpf} represent the electromagnetic power variation corresponding to the droop control and the virtual inertia control of the wind turbine, respectively.

According to Equation (9), the rotor speed variation in the wind turbine participating in primary frequency regulation is obtained as

$$\Delta\omega_r(s) = \frac{\Delta P_e(s)}{2H_w s - K_{cp}} = \Delta\omega_{rK_{df}}(s) + \Delta\omega_{rK_{pf}}(s) \quad (10)$$

where

$$\begin{aligned}\Delta\omega_{rK_{df}}(s) = &\frac{RP_{step}K_{df}}{2H_w(DR+K_m+RK_{df})}\times\\ &\left(\frac{A_0}{s}+\frac{A_1}{s-K}+\frac{Bs+C}{s^2+2\zeta\omega_n s+\omega_n^2}\right)\end{aligned} \quad (11a)$$

$$\begin{aligned}\Delta\omega_{rK_{pf}}(s) = &\frac{RP_{step}K_{pf}}{2H_w(DR+K_m+RK_{df})}\times\\ &\left(\frac{A_1'}{s-K}+\frac{B's+C'}{s^2+2\zeta\omega_n s+\omega_n^2}\right)\end{aligned} \quad (11b)$$

The time domain expression of the rotor speed variation in the wind turbine obtained from the simultaneous Equation (11a,b) is

$$\Delta\omega_r(t) = \Delta\omega_{rK_{df}}(t) + \Delta\omega_{rK_{pf}}(t) \quad (12)$$

The detailed derivation processes of each coefficient in Equations (11a) to (12) are presented in Appendix A.

From Equation (12), the rotor speed variation $\Delta\omega_{roff}$ corresponding to the exiting time t_{off} of the primary frequency regulation of the wind turbine can be calculated. $\Delta\omega_{roff}$ is affected by K_{df}, K_{pf}, and t_{off} simultaneously, and in turn affects the range of values for t_{off}. This parameter t_{off} is essential for the design of the optimization method for the control parameters of wind turbines' primary frequency regulation.

4. Dynamic Frequency Calculation Considering Wind Turbines Participating in Primary Frequency Regulation

4.1. Calculation of the Primary Frequency Drop and Its Minimum Value

When $0 < \zeta < 1$, the time domain expression of the system frequency variation Δf can be obtained from Equation (7) as

$$\begin{aligned}\Delta f(t) &= \Delta\omega(t) \\ &= \frac{RP_{step}}{DR+K_m+RK_{df}}[1+\alpha\cdot\exp(-\zeta\omega_n t)\cdot\sin(\omega_d t+\phi)]\end{aligned} \tag{13}$$

where

$$\begin{cases}\alpha=\sqrt{1+\frac{(T_R\omega_n^2-\zeta\omega_n)^2}{\omega_d^2}} \\ \omega_d=\omega_n\sqrt{1-\zeta^2} \\ \phi=\arctan\left(\frac{\sqrt{1-\zeta^2}}{\zeta-T_R\omega_n}\right)\end{cases} \tag{14}$$

Equation (13) has multiple moments when the derivative is zero, as

$$t_z=\frac{n\pi-\phi_1}{\omega_d} \tag{15}$$

Substituting Equation (15) into Equation (13), the system frequency variation corresponding to these moments is

$$\Delta f_{ex}(n)=\frac{RP_{step}}{DR+K_m+RK_{df}}\left[1+\alpha\cdot\exp(\frac{-\zeta(n\pi-\phi_1)}{\sqrt{1-\zeta^2}})\cdot\sin(n\pi-\phi_2)\right] \tag{16}$$

where

$$\begin{cases}\phi_2=\arctan\left(\frac{-\sqrt{1-\zeta^2}}{\zeta}\right) \\ \phi_1=\phi+\phi_2\end{cases} \tag{17}$$

It can be seen from Appendix B that when $n = 1$, the system frequency reaches the minimum value, and the time corresponding to the minimum value of the system frequency is

$$t_{min}=\frac{(\pi-\phi_1)}{\omega_d} \tag{18}$$

Substituting Equation (18) into Equation (16) and solving it, when $0 < \zeta < 1$, the minimum value of the frequency variation is

$$\Delta f_{min}=\frac{RP_{step}}{DR+K_m+RK_{df}}\cdot\left[1+\alpha\cdot\exp\left(\frac{-\zeta(\pi-\phi_1)}{\sqrt{1-\zeta^2}}\right)\cdot\sin(\pi-\phi_2)\right] \tag{19}$$

When $\zeta > 1$, the time domain expression of Δf is

$$\begin{aligned}\Delta f(t) = \ & \Delta\omega(t)=\frac{RP_{step}}{DR+K_m+RK_{df}}\cdot\left\{1-\frac{1}{2\sqrt{\zeta^2-1}}\cdot\right. \\ & \left[\left(\frac{1}{\zeta-\sqrt{\zeta^2-1}}-T_R\omega_n\right)\cdot\exp(-(\zeta-\sqrt{\zeta^2-1})\omega_n t\right)- \\ & \left.\left.\left(\frac{1}{\zeta+\sqrt{\zeta^2-1}}-T_R\omega_n\right)\cdot\exp(-(\zeta+\sqrt{\zeta^2-1})\omega_n t)\right]\right\}\end{aligned} \tag{20}$$

Then, the time corresponding to the minimum Δf is

$$t_{\min} = \frac{\ln \frac{1-T_R\omega_n\left(\zeta-\sqrt{\zeta^2-1}\right)}{1-T_R\omega_n\left(\zeta+\sqrt{\zeta^2-1}\right)}}{-2\omega_n\sqrt{\zeta^2-1}} \tag{21}$$

The lowest value of the system frequency drops when $\zeta > 1$ is obtained from Equations (20) and (21).

4.2. Calculation of the Secondary Frequency Drop and Its Minimum Value

When the wind turbine exits the primary frequency regulation, the switch O in Figure 2 is open and only traditional generator units in the power system continue to contribute to the primary frequency regulation. The SFR model shown in Figure 3 is established by setting the initial time t_{off} as the new reference point at which the wind turbine exits the primary frequency regulation. In Figure 3, P_{d2} is the unbalanced power of the system at the time t_{off} and $\Delta\omega_{\text{sc}}$ is the rotor speed variation in the traditional generator unit after the wind turbine exits primary frequency regulation.

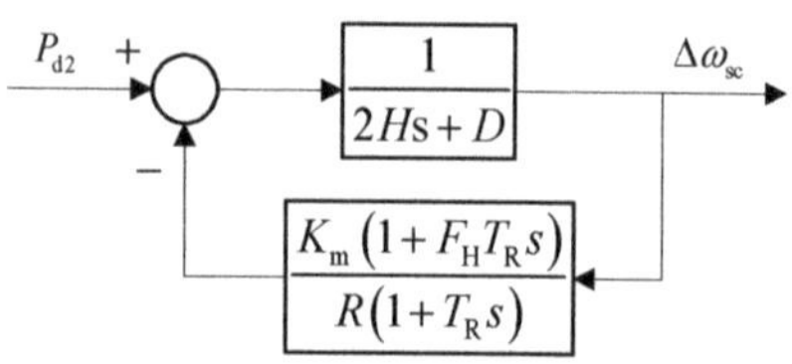

Figure 3. SFR model after wind turbines exit primary frequency regulation.

The ratio of the wind turbine power at the MPPT point to the cube of the wind rotor speed is K_{w}:

$$K_{\text{w}} = n_{\text{w}} \cdot \frac{1}{2}\rho\pi C_{\text{pmax}} \frac{R_{\text{w}}^5}{\lambda_{\text{opt}}^3} \tag{22}$$

where C_{pmax} is the maximum wind power coefficient. λ_{opt} is the optimal tip speed ratio of the wind turbine.

From Equation (22), the difference ΔP_{mppt} between the power at the initial operating point and the power at the MPPT point of the wind turbine at the moment that the wind turbine exits the primary frequency regulation is

$$\Delta P_{\text{mppt}} = K_{\text{w}}[\omega_{\text{r0}}^3 - (\omega_{\text{r0}} + \Delta\omega_{\text{roff}})^3] \tag{23}$$

where $\Delta\omega_{\text{r0}}$ is the wind turbine initial rotor speed.

P_{d2} is still a step function, and the step amount is

$$P_{\text{step2}} = P_{\text{step}} - \Delta P_{\text{mppt}} \tag{24}$$

Figure 1 illustrates that P_{d2} should also include the power component of the wind turbine rotor speed recovery stage, but the recovery time is long, and the electromagnetic power of the wind turbine changes relatively slowly, which has limited influence on the system frequency [9]. In order to simplify the analysis, this paper ignores the influence of the wind turbine rotor speed recovery stage on the secondary frequency drop.

Decompose the SFR model in Figure 3, as shown in Figure 4. The SFR model, following decomposition, comprises two state variables, x_1 and x_2.

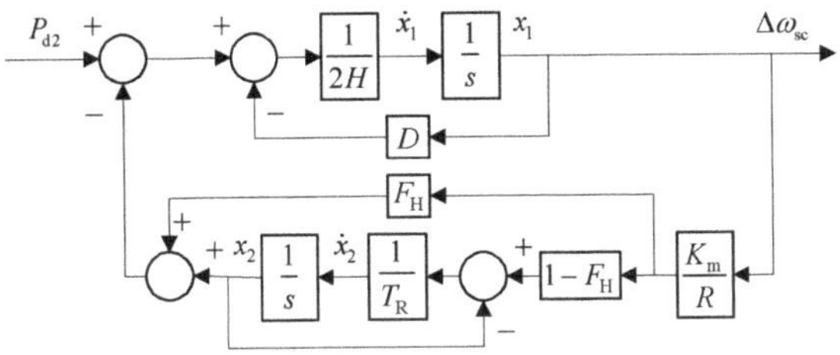

Figure 4. Decomposed SFR model.

The expression in state space of the decomposed SFR model in Figure 4 is

$$\begin{cases} \dot{\boldsymbol{x}} = \boldsymbol{A}_{\mathbf{sc}}\boldsymbol{x} + \boldsymbol{B}_{\mathbf{sc}}\boldsymbol{u} \\ \boldsymbol{y} = \boldsymbol{C}_{\mathbf{sc}}\boldsymbol{x} + \boldsymbol{D}_{\mathrm{sc}}\boldsymbol{u} \end{cases} \tag{25}$$

where

$$\begin{cases} \boldsymbol{A}_{\mathbf{sc}} = \begin{bmatrix} -\frac{K_m F_H + DR}{2HR} & -\frac{1}{2H} \\ \frac{K_m(1-F_H)}{RT_R} & -\frac{1}{T_R} \end{bmatrix} \\ \boldsymbol{B}_{\mathbf{sc}} = \begin{bmatrix} \frac{1}{2H} \\ 0 \end{bmatrix} \\ \boldsymbol{C}_{\mathbf{sc}} = \begin{bmatrix} 1 & 0 \end{bmatrix} \\ \boldsymbol{D}_{\mathrm{sc}} = 0 \end{cases} \tag{26}$$

Assuming that the initial value of the state vector is $\boldsymbol{x}(0)$, solve Equation (25).

$$Y(s) = \boldsymbol{C}_{\mathbf{sc}}(s\boldsymbol{I} - \boldsymbol{A}_{\mathbf{sc}})^{-1}[\boldsymbol{x}(0) + \boldsymbol{B}_{\mathbf{sc}}U(s)] \tag{27}$$

where

$$\begin{cases} U(s) = \frac{P_{\mathrm{step2}}}{s} \\ \boldsymbol{x}(0) = \begin{bmatrix} x_1(0) \\ x_2(0) \end{bmatrix} \end{cases} \tag{28}$$

Simplifying Equation (27), the system frequency deviation Δf_{sc} after the wind turbine exits primary frequency regulation is

$$\begin{aligned} \Delta f_{\mathrm{sc}}(s) = \; & \Delta\omega_{\mathrm{sc}}(s) = \frac{RP_{\mathrm{step2}}}{DR+K_m} \cdot \frac{(1+T_R s)\omega_n'^2}{s(s^2+2\zeta'\omega_n' s+\omega_n'^2)} + \\ & \frac{x_1(0)}{T_R} \cdot \frac{1+T_R s}{s^2+2\zeta'\omega_n' s+\omega_n'^2} - \\ & \frac{x_2(0)}{2H} \cdot \frac{1}{s^2+2\zeta'\omega_n' s+\omega_n'^2} \end{aligned} \tag{29}$$

where

$$\begin{cases} \omega_n'^2 = \frac{DR+K_m}{2HRT_R} \\ \zeta' = \frac{2HR+(DR+K_m F_H)T_R}{2(DR+K_m)} \cdot \omega_n' \end{cases} \tag{30}$$

From Equation (29), the time domain expression of Δf_{sc} is

$$\Delta f_{\mathrm{sc}}(t) = \frac{RP_{\mathrm{step2}}}{DR+K_m} + \mathrm{e}^{-\zeta'\omega_n' t}\alpha_x \sin(\omega_d' t + \phi_x) \tag{31}$$

where

$$\begin{cases} \alpha_x = \sqrt{a_x^2 + b_x^2} \\ \phi_x = \arctan(\frac{b_x}{a_x}) \end{cases} \tag{32}$$

where

$$\begin{cases} a_x = \frac{RP_{step2}(T_R\omega_n'^2 - \zeta'\omega_n')}{(DR+K_m)\omega_d'} + \\ \frac{x_1(0)(1-\zeta'\omega_n'T_R)}{T_R\omega_d'} - \frac{x_2(0)}{2H\omega_d'} \\ b_x = -\frac{RP_{step2}}{DR+K_m} + x_1(0) \end{cases} \quad (33)$$

Substituting the moment when the first derivative is zero into Equation (31), the minimum value of the secondary frequency drop is

$$\Delta f_{scmin} = \frac{RP_{step2}}{DR+K_m} + \alpha_x \cdot \exp\left(\frac{-\zeta'(\pi - \phi_{x1})}{\sqrt{1-\zeta'^2}}\right) \cdot \sin(\pi - \phi_{x2}) \quad (34)$$

where

$$\begin{cases} \phi_{x2} = \arctan\left(\frac{-\omega_d'}{\zeta'\omega_n'}\right) \\ \phi_{x1} = \phi_x + \phi_{x2} \\ \omega_d' = \omega_n'\sqrt{1-\zeta'^2} \end{cases} \quad (35)$$

4.3. State Variable Initial Values Calculation

The state variables x_1 and x_2 are the rotor speed variation $\Delta\omega_{sc}$ of the traditional power unit and the output power variation ΔP_{LP} of the low-pressure cylinder of the traditional generator unit, respectively; the initial value is the value at the time t_{off} when the wind turbine exits the primary frequency regulation.

From Figure 2, $\Delta\omega(s)$ and $\Delta P_{LP}(s)$ can be obtained when the wind turbine participates in the primary frequency regulation. The expression of $\Delta\omega(s)$ has been given in Equation (7) and $\Delta P_{LP}(s)$ is

$$\Delta P_{LP}(s) = \frac{P_{step}K_m(1-F_H)}{DR+K_m+RK_{df}} \cdot \frac{\omega_n^2}{s(s^2+2\zeta\omega_n s+\omega_n^2)} \quad (36)$$

When $0 < \zeta < 1$, the time domain expression of $\Delta\omega(t)$ has been given in Equation (13) and $\Delta P_{LP}(t)$ is

$$\Delta P_{LP}(t) = \frac{P_{step}K_m(1-F_H)}{DR+K_m+RK_{df}} \cdot \left[1 - \frac{1}{\sqrt{1-\zeta^2}} \cdot \exp(-\zeta\omega_n t) \cdot \sin(\omega_d t + \phi_{LP})\right] \quad (37)$$

where

$$\phi_{LP} = \arctan\left(\frac{\sqrt{1-\zeta^2}}{\zeta}\right) \quad (38)$$

When $\zeta > 1$, the time domain expression of $\Delta\omega(t)$ has been given in Equation (20) and $\Delta P_{LP}(t)$ is

$$\Delta P_{LP}(t) = \frac{P_{step}K_m(1-F_H)}{DR+K_m+RK_{df}} \cdot \left\{1 - \frac{1}{2\sqrt{\zeta^2-1}} \cdot \left[\frac{\exp(-(\zeta-\sqrt{\zeta^2-1})\omega_n t)}{\zeta-\sqrt{\zeta^2-1}} - \frac{\exp(-(\zeta+\sqrt{\zeta^2-1})\omega_n t)}{\zeta+\sqrt{\zeta^2-1}}\right]\right\} \quad (39)$$

According to the values of ζ, the wind turbine primary frequency regulation exiting time t_{off} is substituted into Equation (13), Equation (37) or Equation (20), Equation (39), and the initial values of the state variables $x_1(0)$ and $x_2(0)$ are obtained.

5. Parameters Optimization of the Wind Turbine Participating in Primary Frequency Regulation

5.1. Optimization Objective and Solution Vector

Taking the minimum values both of the primary and secondary frequency drops when the wind turbine participates in the primary frequency regulation as the optimization objective, a set of multivariate function nonlinear optimization models is constructed.

Discussion on the classification of optimization objective function z is as follows. When the primary frequency drop is more severe, the minimum value of the primary drop should be as high as possible, otherwise the minimum value of the secondary frequency drop should be as high as possible. Therefore, the optimization objective function of the primary frequency regulation control parameters of the wind turbine is

$$\begin{cases} z = \max(\Delta f_{\text{min}}), \Delta f_{\text{min}} \leq \Delta f_{\text{scmin}} \\ z = \max(\Delta f_{\text{scmin}}), \Delta f_{\text{min}} \geq \Delta f_{\text{scmin}} \end{cases} \tag{40}$$

The primary frequency regulation effect of the wind turbine is influenced by three factors: the droop control parameter K_{df}, the virtual inertia control parameter K_{pf} and the exiting time t_{off}. Therefore, the solution vector of the optimization model is

$$\boldsymbol{x_o} = [K_{\text{df}}, K_{\text{pf}}, t_{\text{off}}] \tag{41}$$

5.2. Constraint Conditions

The optimization model of wind turbine primary frequency control parameters contains several equality and inequality constraints.

When $0 < \zeta < 1$, simultaneous Equations (12), (18), (19) and (34), the equality constraints are

$$\begin{cases} \Delta\omega_{\text{roff}} = \Delta\omega_{\text{r}K_{\text{df}}}(t_{\text{off}}) + \Delta\omega_{\text{r}K_{\text{pf}}}(t_{\text{off}}) \\ \Delta f_{\text{min}} = \frac{RP_{\text{step}}}{DR+K_{\text{m}}+RK_{\text{df}}} \cdot \left[1 + \alpha \cdot \exp(\frac{-\zeta(\pi-\phi_1)}{\sqrt{1-\zeta^2}}) \cdot \sin(\pi - \phi_2)\right] \\ t_{\text{min}} = \frac{(\pi-\phi_1)}{\omega_{\text{d}}} \\ \Delta f_{\text{scmin}} = \frac{RP_{\text{step2}}}{DR+K_{\text{m}}} + \alpha_{\text{x}} \cdot \exp(\frac{-\zeta'(\pi-\phi_{\text{x1}})}{\sqrt{1-\zeta'^2}}) \cdot \sin(\pi - \phi_{\text{x2}}) \end{cases} \tag{42}$$

When $\zeta > 1$, simultaneous Equations (12), (20), (21) and (34), the equality constraints are

$$\begin{cases} \Delta\omega_{\text{roff}} = \Delta\omega_{\text{r}K_{\text{df}}}(t_{\text{off}}) + \Delta\omega_{\text{r}K_{\text{pf}}}(t_{\text{off}}) \\ \Delta f(t) = \frac{RP_{\text{step}}}{DR+K_{\text{m}}+RK_{\text{df}}} \cdot \left\{1 - \frac{1}{2\sqrt{\zeta^2-1}} \cdot \right. \\ \quad \left[\left(\frac{1}{\zeta-\sqrt{\zeta^2-1}} - T_{\text{R}}\omega_{\text{n}}\right) \cdot \exp(-(\zeta - \sqrt{\zeta^2-1})\omega_{\text{n}}t)\right. \\ \quad \left.\left.\left(\frac{1}{\zeta+\sqrt{\zeta^2-1}} - T_{\text{R}}\omega_{\text{n}}\right) \cdot \exp(-(\zeta + \sqrt{\zeta^2-1})\omega_{\text{n}}t)\right]\right\} \\ t_{\text{min}} = \frac{\ln\frac{1-T_{\text{R}}\omega_{\text{n}}(\zeta-\sqrt{\zeta^2-1})}{1-T_{\text{R}}\omega_{\text{n}}(\zeta+\sqrt{\zeta^2-1})}}{-2\omega_{\text{n}}\sqrt{\zeta^2-1}} \\ \Delta f_{\text{scmin}} = \frac{RP_{\text{step2}}}{DR+K_{\text{m}}} + \alpha_{\text{x}} \cdot \exp(\frac{-\zeta'(\pi-\phi_{\text{x1}})}{\sqrt{1-\zeta'^2}}) \sin(\pi - \phi_{\text{x2}}) \end{cases} \tag{43}$$

Simultaneous Equations (20) and (21) separately lead to the equational constraints on Δf_{min} for $\zeta > 1$.

In order to verify the correctness of the frequency calculation expressions more intuitively, typical parameters $K_{\mathrm{df}} = 3$, $K_{\mathrm{pf}} = 3$, $t_{\mathrm{off}} = 3$ are taken and substituted into the frequency time domain expressions. Figure 5 compares the frequency curves calculated and simulated. It can be seen that the two frequency curves overlap with each other during the entire simulation time.

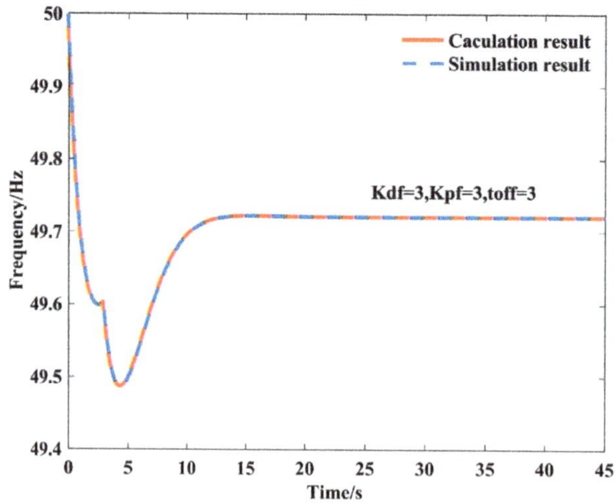

Figure 5. Comparison of frequency curves calculated and simulated.

The inequality constraints of the optimization model are follows.

(1) In order to ensure the stable operation of the wind turbine, the rotor speed of the wind turbine cannot be lower than the operation limit value of 0.7 p.u. and the constraint is

$$-\Delta\omega_{\mathrm{roff}} - (\omega_{\mathrm{r0}} - 0.7) < 0 \tag{44}$$

(2) The traditional generator units should have sufficient time to the system frequency deviation, and the wind turbine should exit the primary frequency regulation after the system frequency drops to the minimum value. So, the constraint is

$$t_{\mathrm{off}} > t_{\mathrm{min}} \tag{45}$$

(3) According to whether the value of ζ is greater than 1 or not, an additional inequality constraint is set up, and one of the following formulas is selected as a constraint during optimization.

$$\begin{cases} 0 < \zeta < 1 \\ \zeta > 1 \end{cases} \tag{46}$$

With the wind turbines participating in the primary frequency regulation, the overall optimal value of the primary and secondary drop of the system frequency is taken as the optimization objective. Figure 6 shows the optimization process.

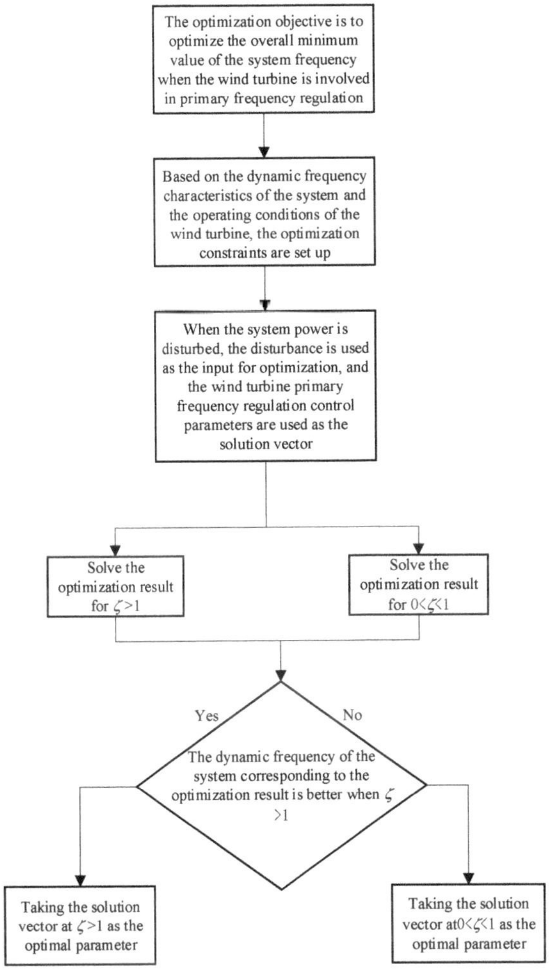

Figure 6. Flow chart of parameters optimization of primary frequency regulation of wind turbines.

6. Case Study

6.1. The Primary Frequency Regulation Effects of the Method Proposed in This Paper

The base value of the test system capacity is 100 MVA, the rated frequency is 50 Hz, and the output power of the conventional generator is 100 MW during normal operation. The base value of the wind turbine blade speed is 1.4273 rad/s, the wind speed is fixed at 12 m/s, the number of wind turbines participating in primary frequency regulation is 5, and the initial operating point power under the wind turbine's single machine equivalence is 0.1185. The angular velocity ratio between the wind turbine rotor and blade is 110, the wind turbine inertia time constant is 1.0185 s, the radius of the wind turbine blade is 40 m, and the values of C_1 to C_5 are 0.22, 116, 5, 12.5, and 0, respectively. The initial rotor speed of the wind turbine is 1.3294 p.u., and the maximum wind power coefficient is 0.4328. The optimal tip speed ratio is 6.325. The air density is 1.25 kg/m^3 and the value of K_{cp} is assumed to be 0.0602. The parameters of the conventional unit are presented in the literature [22].

The function fmincon in MATLAB version: R2020a is used to solve the optimization model. The optimization results of the wind turbine primary frequency regulation control parameters when the system disturbance power is −0.1 p.u. are presented in Table 1, and the simulation results are shown in Figure 7. As the output electromagnetic power when the wind turbine exits primary frequency regulation is less than the initial operating point power, the steady-state value of the frequency when the wind turbine participates in frequency regulation in Figure 7 will be lower than the steady-state value when only the conventional unit is used. As illustrated in Figure 7, regardless of whether ζ is greater than 1 or not, the system frequency curves obtained by solving the two optimization objectives in Equation (41), respectively, overlap completely. This indicates that the optimization process yields the optimal solution when the minimum values of the primary and second frequency drops are equal. Conversely, the system frequency dynamic curve is optimal when $\zeta > 1$, indicating that in comparison to the control parameter of $0 < \zeta < 1$ when K_{pf} is too large and K_{df} is almost zero, the appropriate value of K_{df} has a more pronounced improvement effect on the system frequency.

Table 1. Control parameters and minimum frequency under different optimization results.

	Minimum Frequency/Hz	Droop Control Coefficient	Virtual Inertia Coefficient	Exit Time/s
When $0 < \zeta < 1$	49.61	0.10	28.89	14.80
When $\zeta > 1$	49.67	5.12	6.86	6.33
Only conventional units involved	49.46	/	/	/

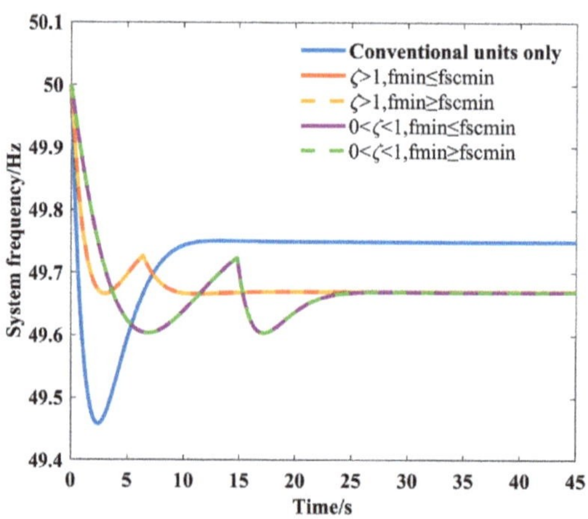

Figure 7. System frequency curves under different optimization results.

Figure 8 illustrates the optimization results of the wind turbine primary frequency regulation control parameters with different disturbance powers. From Figure 8a,b, it can be observed that when the disturbance power is small, the values of K_{df}, K_{pf}, and t_{off} exhibit a tendency to change linearly with the disturbance power. However, they all appear to decline to varying degrees when the disturbance power exceeds −0.4 p.u.

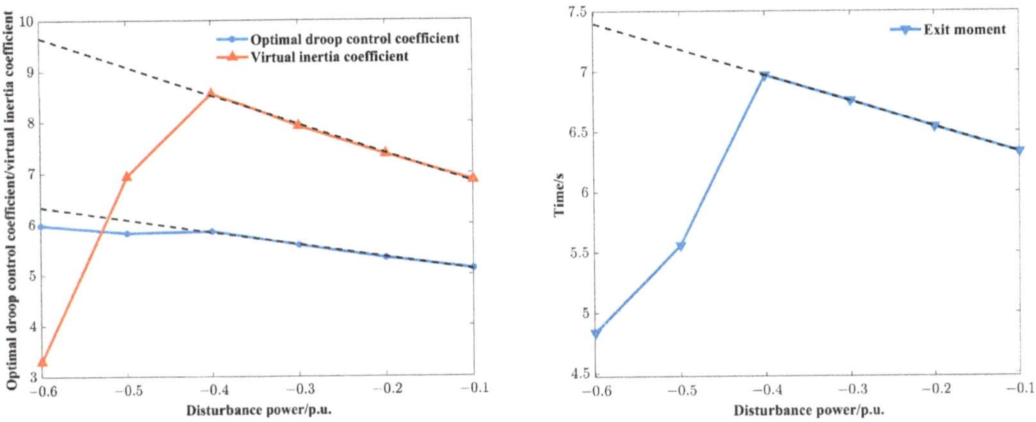

(a) (b)

Figure 8. Optimization results of primary frequency regulation control parameters of wind turbines with different disturbance powers. (**a**) Optimal droop control coefficients/virtual inertia coefficients with different disturbance powers, and (**b**) optimal exit moments with different disturbance powers.

6.2. Comparison of Frequency Regulation Effects with Different Control Parameters and Strategies

The impact of the primary frequency regulation of the wind turbine on the system frequency and the wind turbine's rotor speed is further analyzed when the control parameters take different values. Assuming that the disturbance power is −0.1 per unit, the wind turbine exits the primary frequency regulation at a fixed rotor speed, the rotor speed change is −0.1312 per unit, and the K_{df} and K_{pf} values range from 5 to 7. Table 2 shows a quantitative analysis of the percentage decline in frequency for different control parameters. The system frequency curve of primary frequency regulation of the wind turbine and the wind turbine rotor speed curve are shown in Figure 9a,b, respectively. Figure 9a illustrates that increasing K_{df} and K_{pf} can increase the minimum value of the primary frequency drop, yet simultaneously it will also have a negative impact on the minimum value of the secondary frequency drop. In addition, it can be seen from Figure 9b that there is a positive correlation between the wind turbine's rotor speed change rate and the K_{df}, K_{pf} values. Increasing K_{df} and K_{pf} will accelerate the consumption of wind turbine's rotor kinetic energy, resulting in an early exit time of the wind turbine primary frequency regulation.

Table 2. Comparison of dynamic frequency characteristics of the power system with different control parameters.

	Value of Primary Frequency Drop/Hz	**Value of Secondary Frequency Drop/Hz**	**Exiting Time/s**	**Percentage of Maximum Frequency Drop**
$K_{df} = 5$, $K_{pf} = 5$	49.66	49.67	6.71	0.68%
$K_{df} = 6$, $K_{pf} = 6$	49.68	47.63	5.84	0.74%
$K_{df} = 7$, $K_{pf} = 7$	49.70	49.57	5.26	0.86%
Conventional units only	49.46	/	/	1.08%
The method in this paper	49.67	49.67	6.33	0.66%

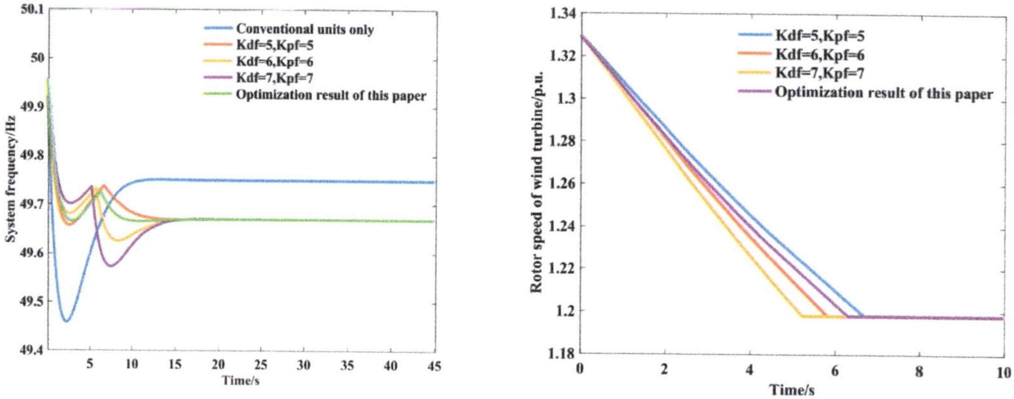

(a) (b)

Figure 9. Comparison of simulation results with different control parameters. (**a**) The system frequency curves with the different control parameters. (**b**) Rotor speed deviation curves of wind turbines with different control parameters.

Figure 10 compares the system frequency curves at different wind turbine primary frequency regulation exit times t_{off} when the disturbance power is -0.1 per unit. The figure illustrates that the minimum frequency with the wind turbine exiting at a fixed time is consistently lower than the optimal solution, regardless of whether the t_{off} is greater or less than the 6.33 s corresponding to the optimization strategy in this paper. This indicates that t_{off} should be employed as part of the parameters optimization process to match the values of K_{df} and K_{pf} to optimize the overall dynamic frequency of the system.

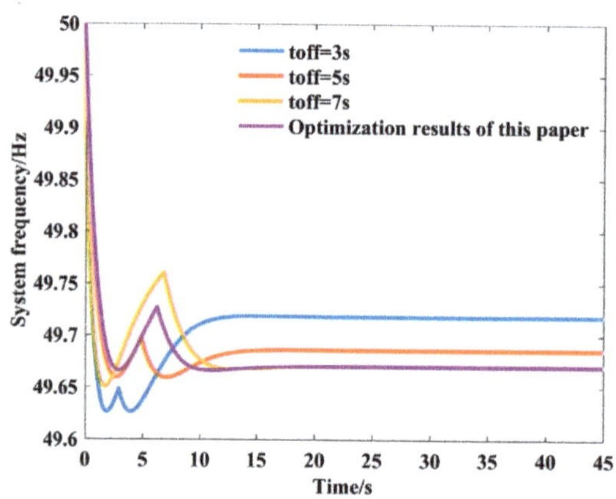

Figure 10. System frequency curves with the different exit moments.

Figure 11 shows the comparison between the method proposed in this paper and the method of preventing the secondary frequency drop in [11]. The method proposed in this paper comprehensively considers the problem of primary and secondary frequency drops to optimize the minimum frequency of the overall process of primary frequency regulation. In contrast, the method in [11] primarily focuses on avoiding or relieving the secondary frequency drop. As illustrated in Figure 11, while the reference method effectively alleviates the issue of the secondary frequency drop, the improvement in the overall frequency minimum is limited and the optimization results are overly conservative. In contrast, although the method proposed in this paper still experiences a secondary frequency drop, the overall dynamic frequency is significantly enhanced in comparison to the traditional method, and the system is less likely to trigger the low-frequency relays. Furthermore, the method proposed in this paper shows a significant reduction in the maximum *RoCoF* when the frequency disturbance occurs, in addition to a delayed system frequency nadir.

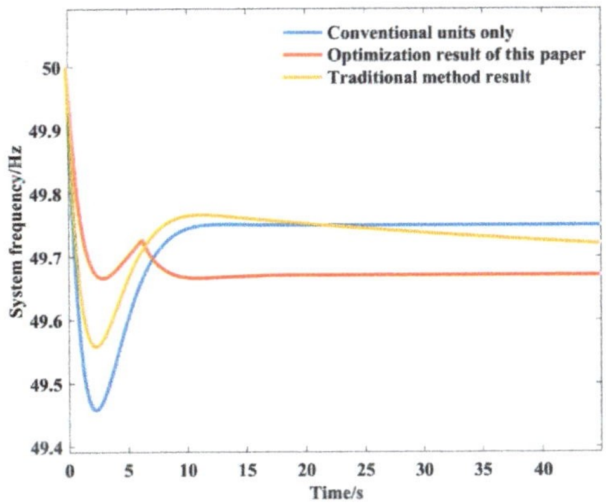

Figure 11. Comparison of optimization results between the proposed method and the method of preventing secondary frequency drop.

7. Conclusions

This study introduces an innovative optimization approach for the control parameters of wind turbine primary frequency regulation, which significantly mitigates both primary and secondary frequency drops. Further studies could explore the scalability of this approach to larger, multi-area power systems, investigate its integration with other renewable energy sources, and develop real-time adaptive control strategies for varying power system conditions. The principal contributions and major findings of this paper are outlined as follows:

(1) The optimization of wind turbine primary frequency regulation control parameters necessitates a precise understanding of the correlation between these parameters and the power system dynamic frequency characteristics. Furthermore, it is essential to consider the interdependent relationship between primary and secondary frequency drops within the power system. Case study analysis demonstrates that the implementation of control parameters which holistically address both primary and secondary frequency drops can substantially enhance the power system dynamic frequency performance.

(2) The optimization results demonstrate that the power system frequency exhibits optimal dynamic characteristics when the minimum values of primary and secondary frequency drops are equivalent. This finding reveals the existence of a game-theoretic relationship between primary and secondary frequency drops.

(3) The optimal configuration of wind turbine primary frequency regulation control parameters is inherently dynamic rather than static. These parameters require coordinated tuning and real-time adjustment in response to both the magnitude of power system disturbances and the operational status of wind turbines.

Author Contributions: Conceptualization, K.L. and Z.C.; methodology, Z.C.; software, Y.G.; validation, X.L.; formal analysis, X.L.; investigation, X.L.; data curation, Z.C.; writing—original draft preparation, K.L.; writing—review and editing, K.L. and Z.C.; visualization, Y.G.; supervision, K.L. All authors have read and agreed to the published version of the manuscript.

Funding: This research received no external funding.

Data Availability Statement: The authors confirm that the data supporting the findings of this study are available within the article.

Conflicts of Interest: The authors declare no conflicts of interest.

Appendix A

The time domain expression for the change in rotor speed of the wind turbine can be solved according to Equation (11a,b).

The coefficients in Equation (11a,b) are

$$\begin{cases} K = \frac{K_{\mathrm{cp}}}{2H_{\mathrm{w}}} \\ A_0 = -\frac{1}{K} \\ A_1 = \frac{-\omega_{\mathrm{n}}^2+4\zeta\omega_{\mathrm{n}}K-T_{\mathrm{R}}\omega_{\mathrm{n}}^2K}{K(K^2+2\zeta\omega_{\mathrm{n}}K-\omega_{\mathrm{n}}^2)} \\ B = \frac{T_{\mathrm{R}}\omega_{\mathrm{n}}^2-2\zeta\omega_{\mathrm{n}}+K}{K^2+2\zeta\omega_{\mathrm{n}}K-\omega_{\mathrm{n}}^2} \\ C = \frac{T_{\mathrm{R}}\omega_{\mathrm{n}}^2(2\zeta\omega_{\mathrm{n}}+K)}{K^2+2\zeta\omega_{\mathrm{n}}K-\omega_{\mathrm{n}}^2} - \frac{4\zeta^2\omega_{\mathrm{n}}^2+2\zeta\omega_{\mathrm{n}}K-\omega_{\mathrm{n}}^2}{K^2+2\zeta\omega_{\mathrm{n}}K-\omega_{\mathrm{n}}^2} \\ A_1' = \frac{\omega_{\mathrm{n}}^2+T_{\mathrm{R}}K\omega_{\mathrm{n}}^2}{K^2+2\zeta\omega_{\mathrm{n}}K+\omega_{\mathrm{n}}^2} \\ B' = \frac{-\omega_{\mathrm{n}}^2-T_{\mathrm{R}}K\omega_{\mathrm{n}}^2}{K^2+2\zeta\omega_{\mathrm{n}}K+\omega_{\mathrm{n}}^2} \\ C' = \frac{T_{\mathrm{R}}\omega_{\mathrm{n}}^4-\omega_{\mathrm{n}}^2(2\zeta\omega_{\mathrm{n}}+K)}{K^2+2\zeta\omega_{\mathrm{n}}K+\omega_{\mathrm{n}}^2} \end{cases} \tag{A1}$$

when $0 < \zeta < 1$,

$$\begin{aligned} \Delta\omega_{\mathrm{r}K_{\mathrm{df}}}(t) = & \frac{RP_{\mathrm{step}}K_{\mathrm{df}}}{2H_{\mathrm{w}}(DR+K_{\mathrm{m}}+RK_{\mathrm{df}})}\Big(A_0 + A_1\cdot\exp(Kt)+ \\ & B\cdot\exp(-\zeta\omega_n t)\cdot\cos(\sqrt{1-\zeta^2}\omega_n t)+ \\ & \frac{C-\zeta\omega_{\mathrm{n}}B}{\sqrt{1-\zeta^2}\omega_{\mathrm{n}}}\cdot\exp(-\zeta\omega_n t)\sin(\sqrt{1-\zeta^2}\omega_n t)\Big) \end{aligned} \tag{A2}$$

$$\begin{aligned} \Delta\omega_{\mathrm{r}K_{\mathrm{pf}}}(t) = & \frac{RP_{\mathrm{step}}K_{\mathrm{pf}}}{2H_{\mathrm{w}}(DR+K_{\mathrm{m}}+RK_{\mathrm{df}})}\Big(A_1'\cdot\exp(Kt)+ \\ & B'\cdot\exp(-\zeta\omega_n t)\cdot\cos(\sqrt{1-\zeta^2}\omega_n t)+ \\ & \frac{C'-\zeta\omega_{\mathrm{n}}B'}{\sqrt{1-\zeta^2}\omega_{\mathrm{n}}}\cdot\exp(-\zeta\omega_n t)\cdot\sin(\sqrt{1-\zeta^2}\omega_n t)\Big) \end{aligned} \tag{A3}$$

when $\zeta > 1$,

$$\begin{aligned} \Delta\omega_{\mathrm{r}K_{\mathrm{df}}}(t) = & \frac{RP_{\mathrm{step}}K_{\mathrm{df}}}{2H_{\mathrm{w}}(DR+K_{\mathrm{m}}+RK_{\mathrm{df}})}\Big(A_0 + A_1\cdot\exp(Kt)+ \\ & \frac{B(\zeta-\sqrt{\zeta^2-1})\omega_{\mathrm{n}}-C}{-2\omega_{\mathrm{n}}\sqrt{\zeta^2-1}}\cdot\exp(-(\zeta-\sqrt{\zeta^2-1})\omega_n t)+ \\ & \frac{C-B(\zeta+\sqrt{\zeta^2-1})\omega_{\mathrm{n}}}{-2\omega_{\mathrm{n}}\sqrt{\zeta^2-1}}\cdot\exp(-(\zeta+\sqrt{\zeta^2-1})\omega_n t)\Big) \end{aligned} \tag{A4}$$

$$\begin{aligned} \Delta\omega_{\mathrm{r}K_{\mathrm{pf}}}(t) = & \frac{RP_{\mathrm{step}}K_{\mathrm{pf}}}{2H_{\mathrm{w}}(DR+K_{\mathrm{m}}+RK_{\mathrm{df}})}\Big(A_1'\cdot\exp(Kt)+ \\ & \frac{B'(\zeta-\sqrt{\zeta^2-1})\omega_{\mathrm{n}}-C'}{-2\omega_{\mathrm{n}}\sqrt{\zeta^2-1}}\cdot\exp(-(\zeta-\sqrt{\zeta^2-1})\omega_n t)+ \\ & \frac{C'-B'(\zeta+\sqrt{\zeta^2-1})\omega_{\mathrm{n}}}{-2\omega_{\mathrm{n}}\sqrt{\zeta^2-1}}\cdot\exp(-(\zeta+\sqrt{\zeta^2-1})\omega_n t)\Big) \end{aligned} \tag{A5}$$

Appendix B

From Equations (13)–(17), it can be observed that there are multiple points with zero derivatives for the primary frequency drop. Among these, the maximum frequency offset $\Delta f_{\min}$, and its corresponding moment $t_{\min}$, are of particular interest in the context of the system frequency stability.

From Equation (16), n only affects the second half of the equation. Consequently, the proportional function can be established as

$$h(n) = \frac{\left|\exp(\frac{-\zeta(n\pi-\phi_1)}{\sqrt{1-\zeta^2}})\cdot\sin(n\pi-\phi_2)\right|}{\left|\exp(\frac{-\zeta(\pi-\phi_1)}{\sqrt{1-\zeta^2}})\cdot\sin(\pi-\phi_2)\right|} = \exp(\frac{-\zeta\pi(n-1)}{\sqrt{1-\zeta^2}}) \leq 1 \tag{A6}$$

The function $h(n)$ is monotonically decreasing; thus, the system frequency reaches its lowest value at $n = 1$.

References

1. Wang, R.; Qin, S.; Bao, W.; Hou, A.; Ying, Y.; Ding, L. Configuration and control strategy for an integrated system of wind turbine generator and supercapacitor to provide frequency support. *Int. J. Electr. Power Energy Syst.* **2023**, *154*, 109456. [CrossRef]
2. Messasma, C.; Barakat, A.; Chouaba, S.E.; Sari, B. PV system frequency regulation employing a new power reserve control approach and a hybrid inertial response. *Electr. Power Syst. Res.* **2023**, *223*, 109556. [CrossRef]
3. Hatziargyriou, N.; Milanovic, J.; Rahmann, C.; Ajjarapu, V.; Canizares, C.; Erlich, I.; Hill, D.; Hiskens, I.; Kamwa, I.; Pal, B.; et al. Definition and Classification of Power System Stability—Revisited & Extended. *IEEE Trans. Power Syst.* **2021**, *36*, 3271–3281.
4. Morren, J.; Haan, S.W.H.D.; Kling, W.L.; Ferreira, J.A. Wind turbines emulating inertia and supporting primary frequency control. *IEEE Trans. Power Syst.* **2006**, *21*, 433–434. [CrossRef]
5. Chen, C.; Su, Y.; Yang, T.; Huang, Z. Virtual inertia coordination control strategy of DFIG-based wind turbine for improved grid frequency response ability. *Electr. Power Syst. Res.* **2023**, *216*, 109076. [CrossRef]
6. Boyle, J.; Littler, T.; Muyeen, S.M.; Foley, A.M. An alternative frequency-droop scheme for wind turbines that provide primary frequency regulation via rotor speed control. *Int. J. Electr. Power Energy Syst.* **2021**, *133*, 107219. [CrossRef]
7. Wang, S.; Tomsovic, K. A Novel Active Power Control Framework for Wind Turbine Generators to Improve Frequency Response. *IEEE Trans. Power Syst.* **2018**, *33*, 6579–6589. [CrossRef]
8. Morren, J.; Pierik, J.; de Haan, S.W.H. Inertial response of variable speed wind turbines. *Electr. Power Syst. Res.* **2006**, *76*, 980–987. [CrossRef]
9. Bao, W.; Ding, L.; Liu, Z.; Zhu, G.; Kheshti, M.; Wu, Q.; Terzija, V. Analytically derived fixed termination time for stepwise inertial control of wind turbines—Part I: Analytical derivation. *Int. J. Electr. Power Energy Syst.* **2020**, *121*, 106120. [CrossRef]
10. Kang, M.; Muljadi, E.; Hur, K.; Kang, Y.C. Stable Adaptive Inertial Control of a Doubly-Fed Induction Generator. *IEEE Trans. Smart Grid* **2016**, *7*, 2971–2979. [CrossRef]
11. Kang, M.; Kim, K.; Muljadi, E.; Park, J.W.; Kang, Y.C. Frequency Control Support of a Doubly-Fed Induction Generator Based on the Torque Limit. *IEEE Trans. Power Syst.* **2016**, *31*, 4575–4583. [CrossRef]
12. Sun, M.; Min, Y.; Xiong, X.; Chen, L.; Zhao, L.; Feng, Y.; Wang, B. Practical Realization of Optimal Auxiliary Frequency Control Strategy of Wind Turbine Generator. *J. Mod. Power Syst. Clean Energy* **2022**, *10*, 617–626. [CrossRef]
13. Meng, L.; Qin, C.; Zeng, Y.; Sun, B., Chen, Q.; Chen, W. A two-stage frequency response method for DFIGs under variable wind speeds. *Electr. Power Syst. Res.* **2023**, *225*, 109813. [CrossRef]
14. Wu, Y.K.; Yang, W.H.; Hu, Y.L.; Dzung, P.Q. Frequency Regulation at a Wind Farm Using Time-Varying Inertia and Droop Controls. *IEEE Trans. Ind. Appl.* **2019**, *55*, 213–224. [CrossRef]
15. Gao, H.; Zhang, F.; Ding, L.; Cornélusse, B.; Zhang, G.; Salimu, A. Multi-segment droop control and optimal parameter setting strategy of wind turbine for frequency regulation. *Int. J. Electr. Power Energy Syst.* **2024**, *158*, 109968. [CrossRef]
16. Nie, Y.; Liu, J.; Gao, L.; Wu, Y.; Li, Z. Nonlinear rotor kinetic energy control strategy of DFIG-based wind turbine participating in grid frequency regulation. *Electr. Power Syst. Res.* **2023**, *223*, 109678. [CrossRef]
17. Sun, M.; Min, Y.; Chen, L.; Hou, K.; Xia, D.; Mao, H. Optimal auxiliary frequency control of wind turbine generators and coordination with synchronous generators. *CSEE J. Power Energy Syst.* **2021**, *7*, 78–85.
18. Lin, C.H.; Wu, Y.K. Coordinated Frequency Control Strategy for VSC-HVDC-Connected Wind Farm and Battery Energy Storage System. *IEEE Trans. Ind. Appl.* **2023**, *59*, 5314–5328. [CrossRef]

19. Xiong, L.; Yang, S.; Huang, S.; He, D.; Li, P.; Khan, M.W.; Wang, J. Optimal Allocation of Energy Storage System in DFIG Wind Farms for Frequency Support Considering Wake Effect. *IEEE Trans. Power Syst.* **2022**, *37*, 2097–2112. [CrossRef]
20. Rahimi, T.; Ding, L.; Kheshti, M.; Faraji, R.; Guerrero, J.M.; Tinajero, G.D.A. Inertia Response Coordination Strategy of Wind Generators and Hybrid Energy Storage and Operation Cost-Based Multi-Objective Optimizing of Frequency Control Parameters. *IEEE Access* **2021**, *9*, 74684–74702. [CrossRef]
21. Xia, Y.; Ahmed, K.H.; Williams, B.W. Wind Turbine Power Coefficient Analysis of a New Maximum Power Point Tracking Technique. *IEEE Trans. Ind. Electron.* **2013**, *60*, 1122–1132. [CrossRef]
22. Anderson, P.M.; Mirheydar, M. A low-order system frequency response model. *IEEE Trans. Power Syst.* **1990**, *5*, 720–729. [CrossRef]

Disclaimer/Publisher's Note: The statements, opinions and data contained in all publications are solely those of the individual author(s) and contributor(s) and not of MDPI and/or the editor(s). MDPI and/or the editor(s) disclaim responsibility for any injury to people or property resulting from any ideas, methods, instructions or products referred to in the content.

energies

MDPI

Article

Effective Customization of Evolutionary Algorithm-Based Energy Management System Optimization for Improved Battery Management in Microgrids

Alessandro Niccolai, Silvia Trimarchi *, Lisa Francesca Barbazza, Alessandro Gandelli, Riccardo Zich, Francesco Grimaccia and Sonia Leva

Department of Energy, Politecnico di Milano, Via Lambruschini 4, 20156 Milan, Italy;
alessandro.niccolai@polimi.it (A.N.); lisafrancesca.barbazza@mail.polimi.it (L.F.B.);
alessandro.gandelli@polimi.it (A.G.); riccardo.zich@polimi.it (R.Z.); francesco.grimaccia@polimi.it (F.G.);
sonia.leva@polimi.it (S.L.)
* Correspondence: silvia.trimarchi@polimi.it

Abstract: The growing penetration of renewable energy sources into electricity grids, along with the problems linked to the electrification of rural areas, has drawn more attention to the development of microgrids. Their Energy Management Systems (EMSs) can be based on evolutionary optimization algorithms to identify efficient scheduling plans and improve performance. In this paper, a new approach based on evolutionary algorithms (EAs) is designed, implemented, and tested on a real microgrid architecture to evaluate its effectiveness. The proposed approach effectively combines heuristic information with the optimization capabilities of EAs, achieving excellent results with reasonable computational effort. The proposed system is highly flexible, making it applicable to different network architectures and various objective functions. In this work, the optimization algorithm directly manages the microgrid Energy Management System, allowing for a large number of degrees of freedom that can be exploited to achieve highly competitive solutions. This method was compared with a standard scheduling approach, and an average improvement of 11.87% in fuel consumption was achieved. After analyzing the differences between the solutions obtained, the importance of the features introduced with this new approach was demonstrated.

Keywords: microgrid; computational intelligence; energy management systems; evolutionary optimization; battery management systems

1. Introduction

Microgrids are playing an increasingly important role in modern energy systems, especially in integrating renewable energy sources (RES), batteries, and Electric Vehicles (EVs) [1]. A microgrid is a localized energy system that can operate independently (islanded mode) or in conjunction with the main grid (on-grid mode), and often includes a range of renewable and non-renewable energy sources, energy storage, and loads. Microgrids can address the intermittent nature of RES, such as solar and wind, by facilitating local energy storage and smart distribution, thereby improving grid reliability and resilience [2]. Thus, the presence of batteries is particularly important, ensuring better RES exploitation during periods of low generation or high demand. This ability to store and dynamically deploy energy not only increases the flexibility and efficiency of microgrids but also supports grid stability by reducing reliance on external power during peak periods. In addition, with the rise of electric vehicles placing new demands on energy infrastructure, microgrids with battery storage offer a means to balance and optimize energy flows [3,4]. Microgrids are important for advancing rural electrification in the developing world, where access to centralized power grids is often limited or non-existent. In addition, they reduce reliance on costly and polluting diesel generators, promoting a cleaner and more self-sufficient energy model that empowers communities and supports economic development. As a result, microgrids provide a scalable, flexible, and cost-effective solution for sustainable rural electrification in the developing world [5].

Energy Management Systems (EMSs) are a fundamental component in the operation of microgrids, optimizing the balance between energy generation, storage, and consumption to ensure efficiency and reliability [6]. As microgrids increasingly incorporate renewable energy sources such as solar and wind, they face challenges in managing variable power generation and fluctuating energy demand. An effective EMS is critical in managing this variability, as it dynamically allocates energy resources based on real-time data, forecasting, and demand response strategies [7]. In this way, an EMS minimizes energy waste, prevents overloading, and maximizes the use of locally generated renewable energy while ensuring stability within the microgrid. In addition, an EMS can intelligently control energy storage systems to store excess energy during off-peak periods and release it during peak periods, improving overall resilience and economic performance [8].

The application of Computational Intelligence (CI) in Energy Management Systems (EMSs) for microgrids has become increasingly important, providing advanced methods for optimizing energy distribution, storage, and demand response [9]. CI techniques, including machine learning, fuzzy logic, and evolutionary algorithms, enable EMSs to process large amounts of real-time data, predict fluctuations in renewable generation and loads [10,11], and adjust to fluctuating demand with high accuracy [12,13], thereby improving economic performance [14].

Focusing on evolutionary optimization algorithms, their landscape is extremely varied and wide, in terms of both implementations and applications. In the engineering field, the most frequently employed algorithms are Particle Swarm Optimization (PSO) and Genetic Algorithms (GAs), which currently represent the traditional benchmarks in the development and comparison of more innovative EAs [15,16]. On the other hand, although the Differential Evolution (DE) algorithm is less frequently employed, it usually outperforms GAs and PSO in engineering applications [17,18]. Finally, Biogeography-Based Optimization (BBO) and Social Network Optimization (SNO) represent two noteworthy EAs that have been developed more recently and shown promising performance in the field [19,20].

Generally, evolutionary algorithms are applied to microgrid EMSs with intermediate layers, such as fuzzy inference systems. These methods effectively reduce the number of design variables; however, they can lead to suboptimal solutions. The direct application of EAs to control microgrid EMSs is generally not used in industrial practice because the high number of design variables and constraints often makes these algorithms slow and unreliable.

Aiming to address these issues, in this paper, a new design variable coding technique is developed to reduce problem complexity without affecting the capability of the algorithm to find optimal and competitive solutions. The proposed approach exploits the power balance constraint and provides the possibility of deterministically defining the optimal setpoint of the dispatchable generators to reduce the number of design variables. Moreover, among all the possible choices of design variables, the most effective one is selected. The methodology is applied to a real microgrid architecture to evaluate its effectiveness.

This paper makes the following key contributions:

- Proposing a flexible framework that can be integrated into nearly any evolutionary algorithm, facilitating the incorporation of more efficient algorithms into the structure.
- Identifying the minimal set of design variables that strikes a balance between convergence speed, non-linearity, and problem complexity. This formulation is designed to be easily integrated with PLC controllers of microgrids.
- Offering a straightforward problem formulation that can be easily adapted to other microgrids with varying components and constraints.

The application of EAs in microgrid EMSs presents a good solution for real-world deployment. EAs are particularly well suited for such applications since they enable flexible implementation on both PLCs and cloud-based platforms. This versatility supports a wide range of operational scenarios and infrastructure configurations. Moreover, since energy planning is performed on a day-ahead basis, the computational time required by EAs does not pose significant limitations. In contrast, MILP approaches are less suitable in this context due to their substantially higher computational demands, especially when the problem includes non-linear cost functions or constraints. The computational complexity of MILP formulations tends to scale poorly with problem size and non-linearity, making them less practical for real-time or resource-constrained environments.

The remainder of this paper is structured as follows. In Section 2, the state of the art in microgrid EMSs is reviewed, and the differences with the proposed approach are highlighted. In Section 3, the microgrid optimization problem is described, focusing on the free parameters, performance metrics, and constraints that should be taken into account in the optimization process. In Section 4, the proposed method is described, the free optimization variables are analyzed, and the cost function is defined. In Section 6, the developed methodology is applied to a specific test case and compared with a basic optimization process. Finally, in Section 7, the conclusions are drawn.

2. Related Works

It is possible to distinguish three different categories of EMSs for microgrids. The first category consists of deterministic techniques, in which the scheduling of the microgrid is determined using algorithms like non-linear programming, the interior point method, and Mixed-Integer Linear Programming (MILP). For these algorithms, the optimal solution is guaranteed, but the model must conform to the algorithm's requirements [21].

Rule-based techniques rely on the definition of scheduling rules that can be activated depending on the real-time conditions of the microgrid. These approaches generally do not provide optimal solutions but can easily handle the stochastic nature of load and RES production [22].

MILP is one of the most popular approaches for EMSs found in the literature [23]. In [24], MILP was used to solve the EMS problem for an off-grid microgrid. The MILP is applied for unit commitment, while a second layer manages real-time operations. The use of MILP has several limitations. The first one concerns the linearization of the working curves of the components: to have a good approximation, it is necessary to use many segments, leading to a large number of variables. The rigid structure required by MILP solvers requires a lot of work to implement the EMS for a different microgrid structure. In addition, many MILP solvers are computationally expensive and can hardly be implemented on the control platforms of rural microgrids.

The use of evolutionary algorithms addresses these problems. Indeed, they can handle non-linear problems and are very flexible, since they can be easily integrated with standard control platforms. Furthermore, EAs are often used in the microgrid design process [25].

In some cases, EAs are applied in combination with fuzzy logic to have a complete tool for EMSs. For example, in [26], a Computational Intelligence-based system was proposed for EMSs. In particular, a Hierarchical Genetic Algorithm was used to optimally tune the rule parameters of a fuzzy inference system (FIS). The optimization process aimed both to reduce the complexity of the FIS and to find the optimal set of parameters. This approach was further investigated in [27], in which a Time-of-Use energy pricing policy was considered. The results showed that the optimization process was capable of improving performance with a reduced number of rules. Another improvement of the original system was proposed in [28], in which the FIS controller was evolved into a neuro-fuzzy system, trained using ad hoc clustering algorithms. This approach, compared to other methods optimized by EAs, reduces the training and operating times, resulting in a more flexible system that can be installed on cheaper machines. These approaches are effective in uncertainty management but often provide sub-optimal solutions because the optimization cannot directly manage the EMS.

In the literature, there are some examples of direct applications of EAs to EMSs. In [29], a modified Particle Swarm Optimization (PSO) technique was proposed for scheduling the production of renewable energy sources in the case of load uncertainty. Similarly, in [30], a Moth Flame Optimization technique was used for managing EMSs in a multi-objective context. With respect to the proposed method, in both of these methods, all the generators were considered independent variables, leading to a very large optimization problem.

Despite advances in EMS development for microgrids, key challenges remain. Deterministic methods like MILP and non-linear programming offer optimal solutions but require strict solver constraints and extensive customization for different configurations. In addition, the necessary model linearization increases both complexity and computational load, posing practical limitations. In contrast, rule-based and fuzzy logic methods manage uncertainty well but often fall short of achieving optimal performance, especially when optimization relies on indirect rule tuning. Finally, while evolutionary algorithms have been explored for EMS scheduling under uncertainty, these methods commonly treat each generator or system component as an independent variable. This leads to an exponential increase in the search space, significantly reducing computational efficiency.

3. Problem Formulation

This section describes the microgrid analyzed, highlighting the free management parameters, the performance metrics, and the constraints to be taken into account in the optimization process.

3.1. Microgrid Single-Node Model

A microgrid is an electrical grid that can operate islanded from (off-grid) or connected to (on-grid) the main grid. It generally comprises dispatchable thermal generators, non-programmable renewable energy sources such as photovoltaics and wind, and one or more loads with different absorption profiles. Finally, it can be equipped with an energy storage system [31].

In case the distances between all the network components are relatively small, a single-node representation of the microgrid can be used, disregarding power-flow constraints [32]. Figure 1 shows the power contributions accounted for in the considered microgrid energy model. This model captures a wide range of possible operating conditions, ensuring adaptability during optimization. This flexibility is required due to the relevant stochastic component in evolutionary algorithms, which randomly generates candidate dispatch solutions both at the beginning and during the entire optimization process.

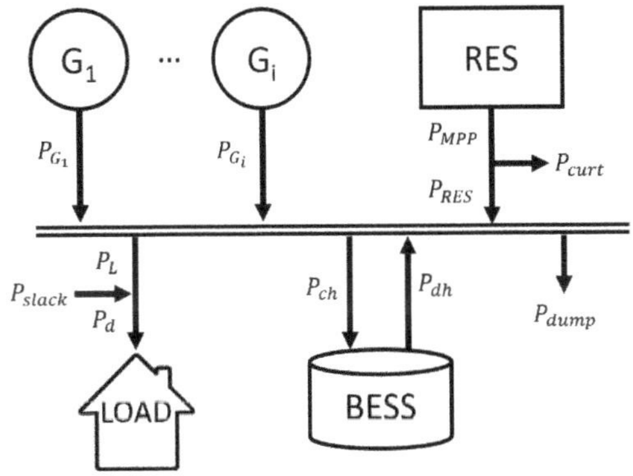

Figure 1. Power flows in a microgrid. The designed procedure can be applied with an arbitrary number of dispatchable generators and is not affected by the nominal size of the components.

The power balance on the common bus of the single-node network, enforced at all the considered time instants, is shown in Equation (1).

$$\sum_{i=1}^{N_G} P_{G_i}(t) + P_{RES}(t) + P_{dh} = P_L(t) + P_{ch}(t) + P_{dump}(t) \tag{1}$$

The first power contribution to the microgrid ($P_{G_i}(t)$) is due to the presence of an arbitrary number N_G of dispatchable generators: their power setpoint can be fully controlled by the EMS, within their technical limits. The operating conditions of these units are defined by their minimum and maximum power and their efficiency curves, linking power production to fuel consumption in a non-linear way.

The contribution of non-programmable RES is determined starting from the forecasted generation potential profiles, i.e., the power produced at the maximum power point ($P_{MPP}(t)$). The microgrid EMS can then curtail the production from renewable sources to obtain the net power fed into the grid P_{RES}, as reported in Equation (2):

$$P_{RES} = P_{MPP} - P_{curt} \tag{2}$$

All the loads present in the network are modeled as a single power absorption point, characterized by an expected load profile ($P_d(t)$). The energy model used, as displayed in Equation (3), accounts for the possibility that part of the load is not covered by production. This case (which is discarded in the optimization process) is modeled using a virtual power generation P_{slack}, i.e., the required power from loads that is not covered by production.

$$P_L = P_d - P_{slack} \tag{3}$$

The storage system can draw ($P_{ch}(t)$) energy and inject ($P_{dh}(t)$) energy into the network, based on the decisions of the EMS. Finally, the model admits the possibility of having excess power generation not absorbed by the loads ($P_{dump}(t)$). This term is also characteristic of non-optimal solutions.

The energy storage system is defined by its nominal capacity (C_N), its minimum and maximum states of charge (SoC) that can be achieved (SoC_{min} and SoC_{max}), and its maximum absorbable and deliverable power ($P_{dh,max}$ and $P_{ch,max}$, respectively) [33].

Finally, the battery is characterized by a charging and discharging efficiency. The operation of the battery energy storage system (BESS) is described in Equations (4) and (5) [34].

$$\Delta \text{SoC}^{+} = \frac{P_{ch} \cdot \Delta t \cdot \eta^{+}}{C_N} \tag{4}$$

$$\Delta \text{SoC}^{-} = \frac{P_{dh} \cdot \Delta t}{\eta^{-} \cdot C_N} \tag{5}$$

where ΔSoC^{+} and ΔSoC^{-} are the state-of-charge variations in charging and discharging, respectively, Δt is the considered time discretization expressed in hours, and η^{+} and η^{-} are the charging and discharging efficiencies.

Given this microgrid model, it is possible to analyze the management variables, performance parameters, and constraints that should be taken into account in the optimization process.

3.2. Microgrid Independent Variables

The microgrid independent variables are the degrees of freedom that can be exploited by the EMS to optimize the microgrid's operating schedule. In particular, for the presented model, they are the power produced by the dispatchable generators, the power absorbed (P_{ch}), and the power delivered (P_{dh}) by the BESS.

The other powers (curtailment, slack, and dump) can be obtained from the free variables identified using the following set of rules. Specifically, they depend on the residual power between production and consumption. The expression for the net power is given by Equation (6), which shows the different terms and contributions employed for its evaluation:

$$P_{net} = \sum_{i=1}^{N_G} P_{G_i} + P_{MPP} + P_{ch} - P_{dh} - P_d \tag{6}$$

If P_{net} is negative, it means that the network is not able to satisfy all the demands of the loads and, consequently, there will be a slack power, as shown in Equation (7):

$$P_{slack} = P_{net}, \qquad P_{net} < 0 \tag{7}$$

Conversely, if P_{net} is positive, production exceeds demand. In this case, renewable generation must be curtailed or surplus energy dumped. As shown in Equations (8) and (9), in the application under analysis, the curtailment is performed first, followed by the dump:

$$P_{curt} = \begin{cases} P_{net}, & 0 < P_{net} < P_{MPP} \\ P_{MPP}, & P_{net} > P_{MPP} \end{cases} \tag{8}$$

$$P_{dump} = P_{net} - P_{curt} \tag{9}$$

This excess power management choice does not affect the optimization process, as the decision to perform curtailment or generation dumping does not have a substantial effect on system performance.

3.3. Constraints

Following the identification of the independent variables, the constraints considered in this study can be systematically analyzed. Notably, the power balance (Equation (1)) is intrinsically maintained through the selection of free variables and the corresponding computation of the power dissipation or slack load.

The constraints of the problem arise from both the operational requirements of the microgrid and the technical limitations of its components.

To avoid trivial solutions where all the power of the loads is slack, it is necessary to impose the first constraint. As shown in Equation (10), the slack is set to zero:

$$P_{slack} = 0\ \text{W} \tag{10}$$

In the case that the size of dispatchable generators is adequate to cover all the loads, it is possible to always satisfy this constraint.

At this point, it is necessary to consider the constraints related to the technical limits of the microgrid components. Starting from dispatchable generators, it is possible to write the constraints on the minimum and maximum power, as shown in Equation (11):

$$P_{G_i} = \begin{cases} 0\ \text{W} & z_i(t) = 0 \\ P_{G_i,min} \leq P_{G_i}(t) \leq P_{G_i,max} & z_i(t) = 1 \end{cases} \tag{11}$$

where $z_i(t)$ is the on/off status of the i-th dispatchable generator, which is given by Equation (12):

$$z_i(t) = \begin{cases} 0 & P_{G_i}(t) = 0 \\ 1 & P_{G_i}(t) > 0 \end{cases} \tag{12}$$

In the case of BESSs, it is necessary to consider both the constraints on the minimum and maximum states of charge and the constraints on the maximum absorbed or supplied power [34]. The set of constraints taken into account is shown in Equation (13):

$$\begin{cases} \text{SoC}_{min} \leq \text{SoC}(t) \leq \text{SoC}_{max} \\ P_{ch}(t) \leq P_{ch,max} \\ P_{dh}(t) \leq P_{dh,max} \end{cases} \tag{13}$$

The last constraint considered in this paper is the spinning reserve requirement, which requires always having an upward generation margin to compensate for potential stochastic fluctuations in loads and non-dispatchable generation [24]. Equation (14) gives the expression for the spinning reserve ($\text{SR}_{av}(t)$). This must be provided by the dispatchable units of the microgrid (e.g., the diesel generators) and the BESS.

$$\text{SR}_{av}(t) = \sum_{i=1}^{N_G} z_i(t) P_{max,i} + \frac{(\text{SoC}(t) - \text{SoC}_{min}) \cdot C_N}{\Delta t} \cdot \eta^- \tag{14}$$

where η^- is the discharging efficiency of the BESS and $z_i(t)$ is the on/off status of the i-th dispatchable generator.

The required spinning reserve can be computed, based on [24], as shown in Equation (15).

$$\text{SR}_{req}(t) = (1 - d_D) P_d(t) - (1 - d_R) P_{MPP}(t) \tag{15}$$

where d_D and d_R quantify the maximum expected sudden demand increase and the PV output decrease, respectively, with respect to the forecasted values.

The formulation of the spinning reserve constraint is given by Equation (16):

$$\text{SR}_{av}(t) \geq \text{SR}_{req}(t) \tag{16}$$

All the presented constraints must be enforced at all time intervals considered; consequently, it can be seen that the optimization problem is highly constrained.

3.4. Performance Parameters

The main goal in the optimization of the microgrid EMS is to reduce operating costs. This can be achieved by minimizing the total fuel consumption. At each time step, the fuel consumption can be calculated as shown in Equation (17):

$$F_i(t) = P_{G_i}(t) \cdot \eta_{G_i}(P_{G_i}) \cdot z_i(t) \tag{17}$$

where $P_{G_i}(t)$ is the power generated by the dispatchable units, $\eta_{G_i}(P_{G_i})$ is the generation efficiency at that power level, and $z_i(t)$ is the on/off status of the generators.

By analyzing this equation, it can be seen that there are two ways to improve system performance: first, by reducing the power contribution from the dispatchable generators, and second, by operating them at load levels with higher efficiency. In general, both of these objectives can be achieved by efficiently managing renewable energy sources and the battery storage system.

4. Method

In this section, the proposed customization method for the evolutionary algorithm-based EMS is described. This method aims to be flexible in terms of both the microgrid model and the selected EAs. Moreover, it should provide effective results. It is based on the proper choice of optimization variables and the exploitation of heuristic constraints to reduce the problem dimensionality. Moreover, the cost function is defined to help the algorithm converge using a proper penalty for infeasible and sub-optimal solutions.

This section is structured as follows. First, the general framework of the method is proposed, and then each step is detailed.

4.1. Optimization Framework

The proposed method is designed, taking into account the selection of optimization variables, the calculation of the physical free problem variables from the optimization variables, and the cost calculation.

Through its specific operators, the optimization algorithm generates a set of candidate solutions (blue blocks in Figure 2). Each candidate solution represents a possible trajectory of the battery energy storage system (BESS) within the microgrid. Based on this information and by utilizing the microgrid model, all relevant power flows are computed, including the power exchanges of the BESS, the setpoints for the dispatchable generators, and the overall system balancing (purple blocks in Figure 2). The balancing process depends on the specific operating scenario considered (green blocks in Figure 2).

The computed power flows are subsequently used by the cost function module (orange blocks in Figure 2) to evaluate constraint violations, apply penalty terms, and estimate fuel consumption. The total cost value is then calculated according to Equation (32). This cost value is fed back into the optimization algorithm, which proceeds to the next iteration, based on the updated evaluations.

The proposed approach differs from the standard methodology in two main aspects. First, it involves the specific selection of optimization variables and a corresponding redefinition of the steps required to compute the power contributions within the system. These steps are designed to minimize the infeasible regions in the search space and reduce the number of optimization variables. Second, a dedicated penalty strategy is introduced to enhance the convergence of the optimization process.

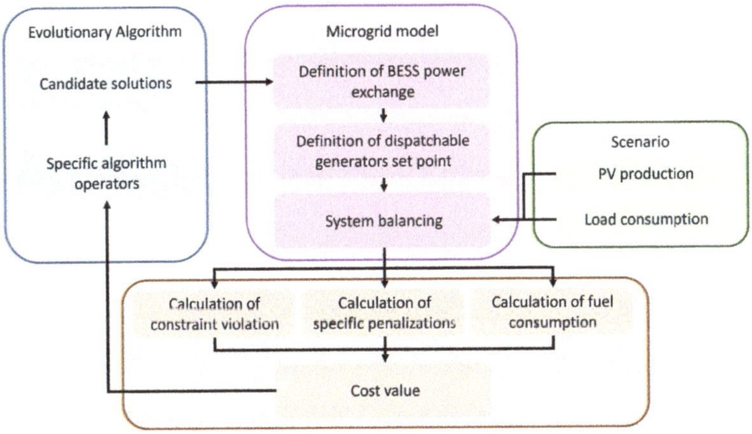

Figure 2. Flowchart of the proposed method. The algorithm uses its operators to create candidate solutions that represent the trajectory of the BESS. The microgrid model computes the power flows that are used in the cost function to close the loop.

4.2. Decodification of Optimization Variables

The first step in the proposed methodology is transforming the optimization variables into the power contributions to the microgrid.

The optimization variables are defined as the BESS state-of-charge values at each time step ($\mathrm{SoC}(t)$). This value should be limited by the minimum and maximum allowable states of charge, as stated in inequality (18):

$$\mathrm{SoC}_{\min} \leq \mathrm{SoC}(t) \leq \mathrm{SoC}_{\max} \tag{18}$$

The corresponding power exchanges are derived by calculating the SoC variation, as shown in Equation (19):

$$\Delta\mathrm{SoC}(t) = \mathrm{SoC}(t) - \mathrm{SoC}(t-1) \tag{19}$$

The power exchanged is then constrained to remain within the BESS operational power limits, i.e., the maximum charging and discharging powers. The corresponding calculations are shown in Equations (20) and (21):

$$P_{dh} = \begin{cases} \max\left(\Delta\mathrm{SoC}(t) \cdot \dfrac{\eta^- C_N}{\Delta t}, -P_{\max,dh}\right) & \Delta\mathrm{SoC}(t) < 0 \\ 0\ \mathrm{W} & \Delta\mathrm{SoC}(t) \geq 0 \end{cases} \tag{20}$$

$$P_{ch} = \begin{cases} 0\ \mathrm{W} & \Delta\mathrm{SoC}(t) \leq 0 \\ \min\left(\Delta\mathrm{SoC}(t) \cdot \dfrac{C_N}{\eta^+ \Delta t}, P_{\max,ch}\right) & \Delta\mathrm{SoC}(t) > 0 \end{cases} \tag{21}$$

These steps are shown in Figure 3. Figure 3a shows a sample profile of the SoC (the blue line); this is defined by the optimizer, and it naturally takes into account the minimum and maximum allowed SoCs (black horizontal lines) from the allowed ranges of the optimization variables. From the computed SoC variations, the BESS power exchanges are then calculated, as shown in Figure 3b; the power limits (black lines) are embedded in this power calculation in order to reduce the constraints that should be considered in the optimization process.

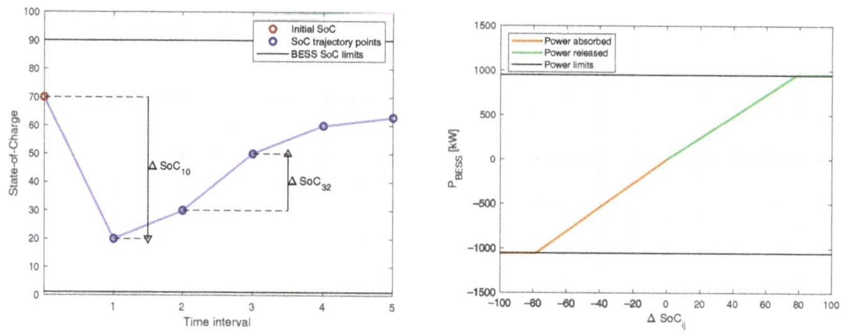

(a) (b)

Figure 3. Steps for defining the P_{BESS} from the optimization variables: (**a**) The optimizer defines the time profile of the state of charge. (**b**) From the SoC variation, the power exchanged by the BESS can be computed, also considering the power constraints.

The power exchanged by the BESS and the scenario information can be used to compute the total power required by the dispatchable generators and, consequently, the power output of each one. The calculation of the required power is performed by exploiting the power balance (Equation (1)), assuming that the slack, dumped, and curtailed power are zero. The formulation used is displayed in Equation (22):

$$P_{tot} = (P_d - P_{MPP}) - P_{BESS} \tag{22}$$

where $P_{BESS} = P_{ch} - P_{dh}$.

Given P_{tot}, the dispatchable generator outputs are determined by solving a simplified optimization problem that minimizes fuel consumption while respecting generator constraints. The definition of the optimization problem is shown in Equation (23):

$$\begin{aligned} \min_{\alpha, z} \quad & \sum_{i=1}^{N_G} F_i(\alpha_i, z_i) \\ \text{s.t.} \quad & \sum_{i=1}^{N_G} z_i \alpha_i P_{tot} \geq P_{tot}\,, \\ & \alpha_i P_{tot} \leq z_i P_{max,i}, \;\; i = 1, \ldots, N_G, \\ & \alpha_i P_{tot} \geq z_i P_{min,i}, \;\; i = 1, \ldots, N_G, \\ & 0 \leq \alpha_i \leq 1, \;\; i = 1, \ldots, N_G \end{aligned} \tag{23}$$

In this minimization problem, the technical limits of the generators are considered. In addition, potential ramp-up and ramp-down limits can be added by properly adjusting the minimum and maximum operating powers of each generator at each time instant, based on its operating load determined in the previous timestep.

If dynamic constraints are not present, this load allocation can be precomputed and excluded from the main EMS optimization loop, reducing overall computational cost. This approach allows the proposed method to minimize the specific consumption at each time step, improving optimization convergence.

The solution of the lower-level minimization problem for the identification of the power contribution of each dispatchable generator can introduce some undesired power contributions (curtailed, dumped, and slack power). These can be calculated using Equations (6)–(9).

The final output of the algorithm is the definition of all the power terms within the microgrid. The last steps required to fully define the optimization method involve managing the constraints and defining the cost function.

4.3. Managing the Constraints and Defining the Cost Function

The proposed selection of the optimization variables allows for a drastic reduction in the number of constraints that must be considered in the optimization process. This is possible because the physical component limits are inherently satisfied.

The two constraints that must be checked are related to the slack power and the spinning reserve. Both are included in the optimization problem using a penalty approach, thus solving an unconstrained problem in which infeasible solutions incur additional cost components.

The main cost component is the total fuel consumption over the analyzed time horizon. In addition to these three cost components, two other penalties for dumped and curtailed power are introduced to improve the algorithm's convergence.

As shown in Equation (24), the resulting cost value comprises five components:

$$C = C_{fuel} + C_{slack} + C_{SR} + C_{dump} + C_{curt} \tag{24}$$

The fuel-related cost is detailed in Equation (25):

$$C_{fuel} = \sum_{t=1}^{N_T} \sum_{i=1}^{N_G} F_i(t) \tag{25}$$

The violation of the slack power constraint is taken into account with the cost component, as shown in Equation (26):

$$C_{slack} = \lambda_1 \cdot \mathcal{H}\left(\sum_{t=1}^{N_T} P_{slack}(t)\right) + \sum_{t=1}^{N_T} P_{slack}(t) \tag{26}$$

where λ_1 is a penalty value and $\mathcal{H}$ is the Heaviside function defined in Equation (27):

$$\mathcal{H}(y) = \begin{cases} 0 & y \leq 0 \\ 1 & y > 0 \end{cases} \tag{27}$$

This definition of the constraint violation cost is introduced because with a proper selection of the value of λ_1 (1.5×10^3 in the analyzed problem), all infeasible solutions have a cost higher than the feasible ones. Moreover, the second penalty term (proportional to the entity of the violation) helps the algorithm converge toward solutions with a lower constraint violation.

The violation of the spinning reserve is considered using a similar approach. The missing spinning reserve can be defined as a function of the difference between the required and available ones, as shown in Equation (28):

$$\mathrm{SR}_{miss}(t) = \begin{cases} \mathrm{SR}_{req}(t) - \mathrm{SR}_{av}(t) & \mathrm{SR}_{req}(t) - \mathrm{SR}_{av}(t) > 0 \\ 0 & \mathrm{SR}_{req}(t) - \mathrm{SR}_{av}(t) \leq 0 \end{cases} \tag{28}$$

Thus, the cost component is formulated in Equation (29):

$$C_{SR} = \lambda_2 \cdot \mathcal{H}\left(\sum_{t=1}^{N_T} \mathrm{SR}_{miss}(t)\right) + \sum_{t=1}^{N_T} \mathrm{SR}_{miss}(t) \tag{29}$$

The curtailment penalty employed, as shown in Equation (30), helps the optimization move toward the optimal solution because it reinforces the benefit of exploiting RES production:

$$C_{curt} = \mu_1 \cdot \sum_{t=1}^{N_T} P_{curt}(t) \tag{30}$$

In this cost component, the step penalty is not introduced because it prevents the correct convergence of the optimization. Moreover, the optimal μ_1 value is very low (1/250 in the analyzed test case).

Finally, the dump power penalty is managed similarly to the curtailment penalty, and its expression is shown in Equation (31):

$$C_{dump} = \mu_2 \cdot \sum_{t=1}^{N_T} P_{dump}(t) \tag{31}$$

In order to achieve proper algorithm convergence, the μ_4 value should be lower than the μ_2 value. In particular, in this work, it is set to 1/500 using a sensitivity analysis.

Once the different terms have been defined, it is possible to present the final and comprehensive formulation of the cost function, as shown in Equation (32). In this way, the different contributions and physical properties that play a role in the objective function can be inferred.

$$\begin{gathered} C = \sum_{t=1}^{N_T} \sum_{i=1}^{N_G} F_i(t) + \lambda_1 \cdot \mathcal{H}\left(\sum_{t=1}^{N_T} P_{slack}(t) \right) + \sum_{t=1}^{N_T} P_{slack}(t) + \lambda_2 \cdot \mathcal{H}\left(\sum_{t=1}^{N_T} \mathrm{SR}_{miss}(t) \right) + \\ \sum_{t=1}^{N_T} \mathrm{SR}_{miss}(t) + \mu_1 \cdot \sum_{t=1}^{N_T} P_{dump}(t) + \mu_2 \cdot \sum_{t=1}^{N_T} P_{curt}(t) \end{gathered} \tag{32}$$

5. Case Study

The proposed methodology was numerically tested on a case study based on the data collected from a real off-grid hybrid microgrid. The required inputs for the simulations are the load consumption profile and the PV-generation potential profile. All the data employed in the study have hourly time resolution.

The microgrid studied can be modeled as detailed in the problem formulation and is composed of two 550 kW peak diesel engines, a 1000 kWp photovoltaic system, a battery energy storage system (BESS) with a maximum capacity of 1440 kWh, and a variable non-programmable load with a maximum power absorption of 928.3 kW. The microgrid is not connected to the main electrical grid. The microgrid scheme is shown in Figure 4.

Some examples of the PV power production profiles and the demand for a sample week in the year are shown in Figure 5.

The microgrid energy storage system has a nominal capacity (C_N) of 1440 kWh; however, a minimum state of charge (SoC) and a maximum value (1% and 90%, respectively) are considered. In terms of power, the BESS is characterized by maximum deliverable power ($P_{dh,max}$ = 1440 kW) and absorbable power ($P_{ch,max}$ = 1440 kW). In the case under analysis, it can be seen that these values do not constitute a particular constraint; however, they are taken into account in the design of the optimization system to ensure maximum flexibility.

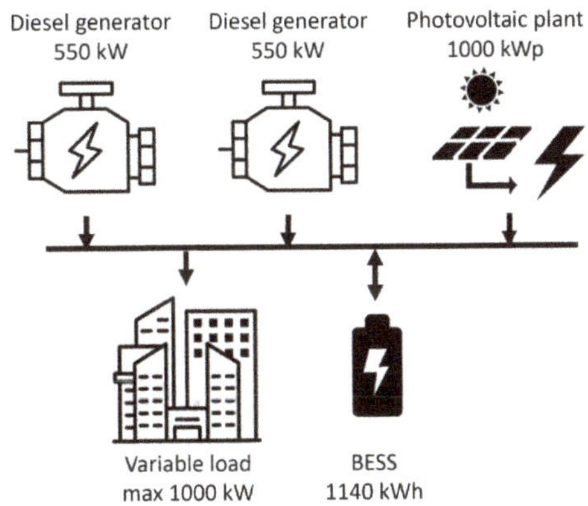

Figure 4. Scheme of the analyzed microgrid.

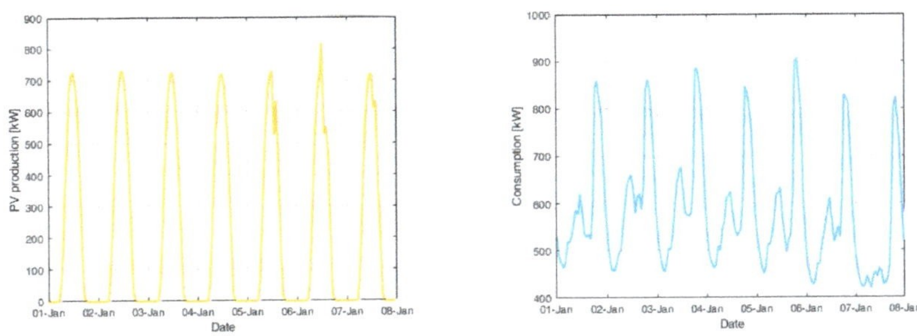

(a) (b)

Figure 5. Examples of (**a**) renewable power production and (**b**) load profile. These values were measured in a real microgrid.

The two diesel engines have a maximum power of 550 kW and a minimum power of 110 kW. Below this threshold, the engine is off ($P_G = 0$ W). Figure 6 shows the details of the consumption curves of the two generators, indicated as G_1 and G_2 in the figure. The left axis shows the consumption as a function of the produced power; the efficiency of the engines is not constant, resulting in a quadratic consumption curve. The right axis shows the specific consumption of the two engines; the reduction in the fuel consumption with the produced power shows that it is convenient to avoid using the engines with low required power.

The optimal power breakdown between the two generators can be calculated a priori by solving the minimization problem in Equation (23). The results of the breakdown as a function of the total required power are shown in Figure 7. Figure 7a shows the power produced by each generator (G_1 in blue and G_2 in orange). Due to the lower specific consumption of G_2, this generator is preferentially used; in particular, if the power production is below its maximum limit, only this engine is activated. Figure 7b shows the total specific consumption of the combination of the two engines (left axis) and the best breakdown value obtained in the minimization problem (right axis). The curve of the specific consumption shows a step corresponding to the ignition of the second engine, which has a lower generation efficiency.

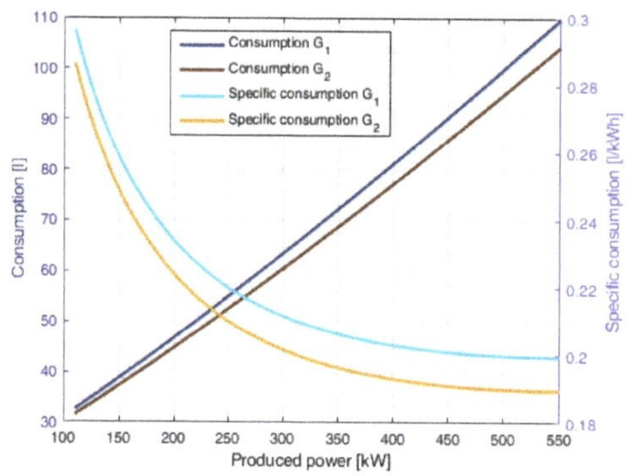

Figure 6. Characteristics of the two engines in the microgrid: the left axis represents the consumption, and the right axis represents the specific consumption.

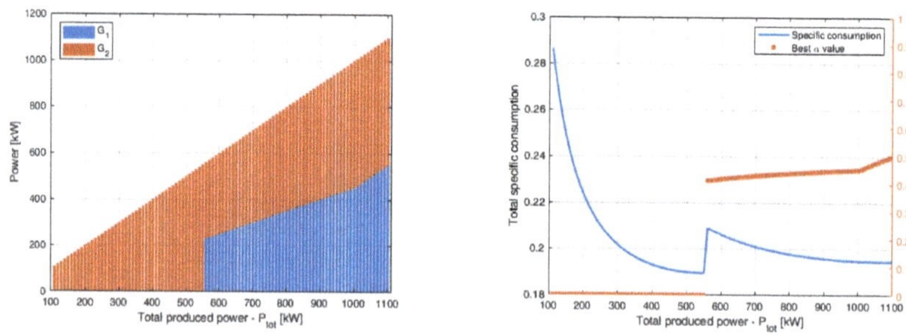

(**a**) (**b**)

Figure 7. Optimal power breakdown between the two dispatchable generators: (**a**) the proportion of power produced by the two generators as a function of the total required power; (**b**) the total specific consumption and the breakdown α value.

6. Results

The proposed method was validated on the described case study, and this section discusses the results of the assessment.

This section is divided into three parts. In the first one, different EAs are applied with the proposed method for optimization on a sample scenario extracted from the described test case. Then, the optimal solution of the best-performing algorithm is analyzed. Finally, the proposed method is compared with a standard optimization procedure on a set of scenarios to assess the robustness of the methodology.

All the analyses were performed on an Intel(R) Core(TM) i9-10900KF 3.70 GHz. The number of objective function calls was used as the termination criterion, as it correlates with total optimization time. It also enables fair comparisons across algorithms with different population sizes. In order to ensure the statistical reliability of the results, several independent trials were performed in all the tests.

Table 1 shows an overview of the tests performed in this section, highlighting the number of cost function calls, the number of independent trials, and the number of scenarios tested.

Table 1. Overview of the tests performed in this section.

Test Case	Cost Function Calls	Independent Trials	Scenarios Tested
Algorithm and parameter selection	100,000	100	1
Best solution analysis	100,000	50	1
Approach comparison	100,000	50	28

6.1. Algorithm and Parameter Selection

To identify the most suitable evolutionary algorithm for the proposed application, a comparative analysis was conducted among five different algorithms: Social Network Optimization (SNO) [35], Differential Evolution (DE) [36], Biogeography-Based Optimization (BBO) [37], Particle Swarm Optimization (PSO), and a Genetic Algorithm (GA) [38].

Table 2 summarizes the working parameters of the algorithms employed. They represent the population size in addition to algorithm-specific parameters, such as the mutation rate for the GA and the inertia for PSO.

Table 2. Parameters of the four tested evolutionary algorithms: (a) Social Network Optimization, (b) Genetic Algorithm, (c) Biogeography-Based Optimization, (d) Particle Swarm Optimization, and (e) the Differential Evolution.

(a) Social Network Optimization		(b) Genetic Algorithm	
Population size	40	Population size	25
Attraction rate	0.64	Crossover probability	0.98
Preserve rate	0.52	Mutation rate	0.1
Linguistic error probability	0.08	Mutation amplitude	0.06
(c) Biogeography-Based Optimization		**(d) Particle Swarm Optimization**	
Population size	20	Population size	20
Immigration coefficient	500	Inertia	0.95
Emigration coefficient	1	Acceleration coefficients	0.45
Mutation rate evolution	0.01	Velocity clamping	0.35
(e) Differential Evolution			
Population size		25	
Crossover probability		0.5	
Differential weight		0.5	

Figure 8 presents the average convergence curves, computed over 100 independent trials, for the evaluated optimization algorithms. The results are displayed on a logarithmic scale to highlight differences in convergence behavior. PSO and the GA exhibited a very rapid initial convergence; however, their progress slowed earlier compared to the other algorithms. In terms of final performance, SNO, DE, BBO, and the GA achieved comparable results. Notably, SNO demonstrated particularly competitive behavior, maintaining superior performance over the majority of the optimization process.

An additional critical aspect in selecting the most appropriate algorithm for EMS applications is the robustness of the optimization process. Figure 9 shows box plots of the results obtained from 100 independent trials for each algorithm. This analysis highlights the strong performance of SNO and DE, which demonstrated not only excellent convergence capabilities but also high consistency across trials, indicating the greater reliability of the solutions produced.

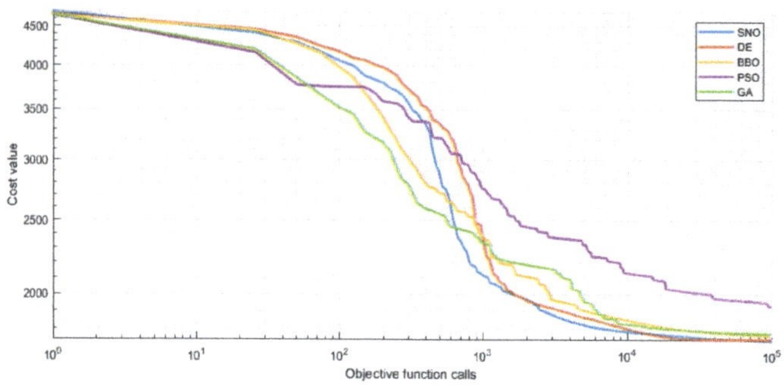

Figure 8. Average convergence curves of all the tested algorithms.

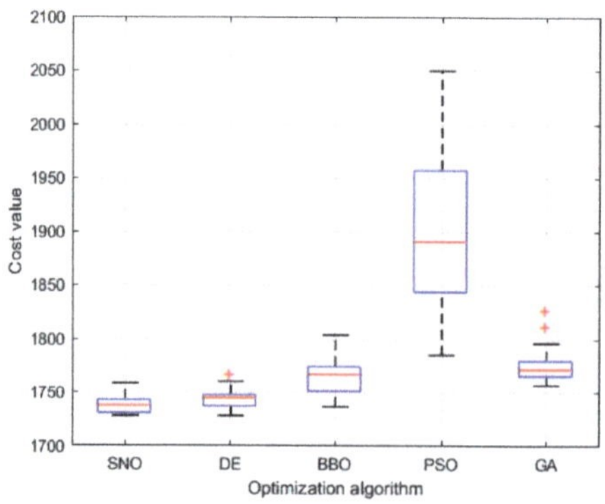

Figure 9. Box plot for statistical assessment of the tested algorithms. The red plus signs represent the outliers.

Finally, since in operation, computational time is very important, Table 3 shows the average computational time required by each of the tested algorithms. Performance was comparable across all the algorithms, and the optimization times were compatible with the specific EMS application. In fact, due to the changing conditions of the loads and weather, it was often necessary to run the optimization several times.

Table 3. Computational time required for the optimization in seconds.

SNO	DE	BBO	PSO	GA
13.61	12.42	14.36	13.44	12.69

From the comparison, the superior performance of the DE and SNO algorithms emerged compared to PSO, the GA, and BBO. Particularly notable was their much higher reliability, inferable from the considerably lower standard deviation values achieved. Therefore, for the application analyzed here, they have more robust optimization capabilities compared to the other three methods. Focusing on the SNO and DE algorithms, Figure 9 shows that SNO achieved both a slightly lower mean value and standard deviation. Therefore, to optimize the current problem, the Social Network Optimization algorithm was selected, and its optimization process and working principles are briefly reported in the next section.

The cost function presented in Equation (32) includes scalarization parameters. Their values were selected based on a sensitivity analysis conducted using the SNO algorithm.

The results of this analysis are shown in Figure 10. While the introduction of these parameters led to a slight improvement in the overall performance of the algorithm, the most significant effect was a noticeable reduction in the dispersion of the results, thereby enhancing the robustness of the optimization process.

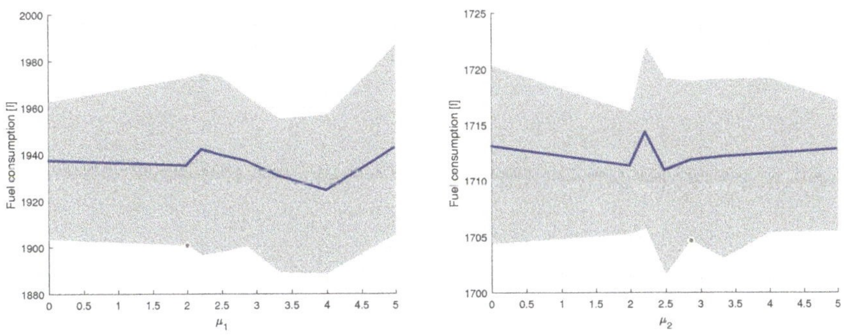

(a) (b)

Figure 10. Sensitivity analysis of the penalty coefficients μ_1 and μ_2, respectively in (**a**) and (**b**): the blue line represents the average fuel consumption, while the gray bands show the dispersion of the results.

6.2. Analysis of SNO Optimization Process

The SNO algorithm is a population-based evolutionary algorithm. It is inspired by the dynamics of social media platforms and the way information circulates within them. In this algorithm, the pool of potential solutions represents users in a virtual social network, imitating how individuals communicate, exchange ideas, and engage with one another online. Given its superior performance, it was selected for the optimization study, and its optimization process was analyzed in more detail.

Figure 11 shows the convergence curves of 50 independent trials of SNO; each trial is represented by a thin gray line, and the thick blue line represents the average convergence. Due to the wide variation in cost values in the optimization, an insert zooms in on the lower cost values. Finally, the histogram on the right of the figure shows the distribution of the final values of each trial.

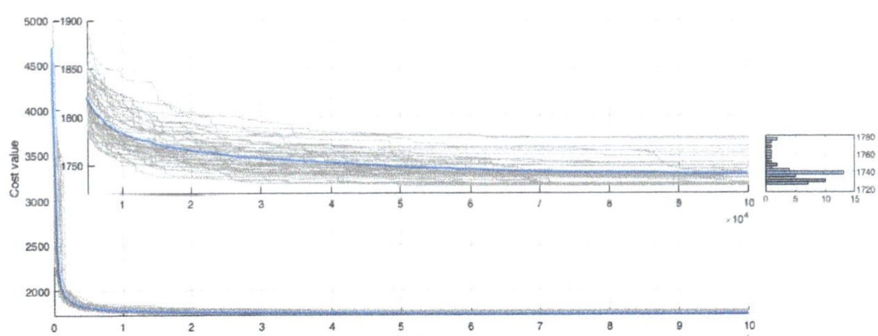

Figure 11. Convergence curves of 50 independent trials of SNO on an optimization scenario. Each gray line represents one trial, while the thick line represents the average convergence. The insert zooms in on lower cost values, while the histogram on the right shows the distribution of the final values.

Analyzing the convergence, a sharp drop can be seen in the early stages of the process, which corresponds to the fulfillment of the constraints, particularly the spinning reserve requirement. Then, the convergence rate decreases because the optimization process enters the exploitation phase to improve the solution quality.

The distribution of the final values is concentrated on low costs, even if some trials do not properly converge. This is due to the intrinsic stochastic nature of evolutionary algorithms. However, the low dispersion indicates good algorithm reliability; thus, only a few independent trials are required in real applications of this approach.

Figure 12 shows the optimal solution found by SNO for the analyzed scenario. Figure 12a represents the power flows in the microgrid; the red line represents the total demand, while the bars represent all the power contributions. The EMS utilizes the first part of the day to charge the battery, drawing energy from both the dispatchable generators and the photovoltaic production. This evolution is evident in Figure 12b, which shows the evolution over time of the state of charge; each scenario analyzed starts with the BESS at its minimum value. Since the remaining energy content in the storage is not in the cost function, the final SoC in the optimal solution is the minimum allowed.

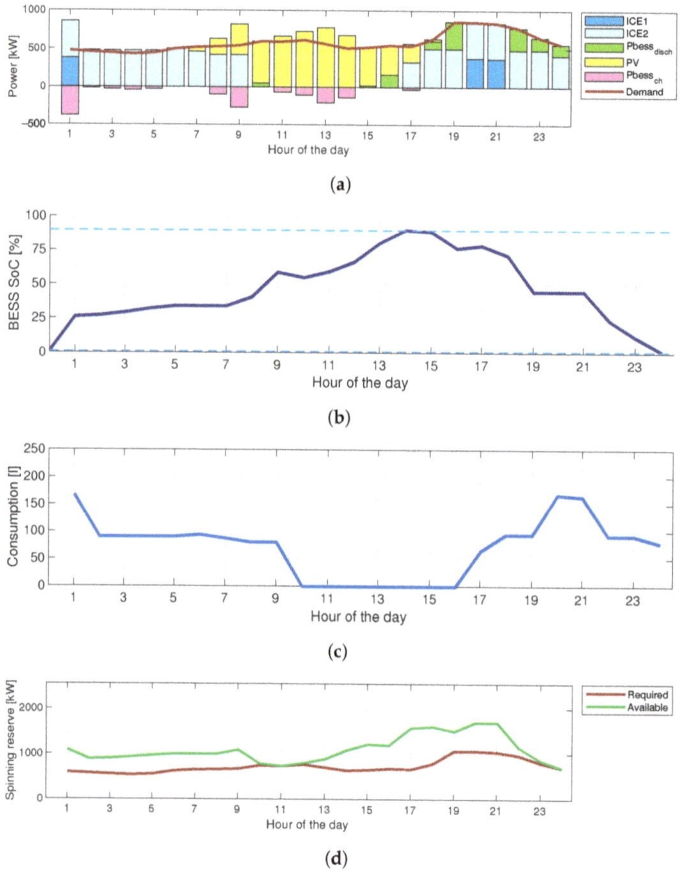

Figure 12. Optimal solution found by SNO: (**a**) Energy flow in each hour of the day; (**b**) Evolution of the BESS SoC; (**c**) Consumption in each hour; (**d**) Spinning reserve.

Figure 12c shows the total consumption of the two dispatchable generators. The main cost component minimized is the integral of this curve.

Finally, Figure 12d shows the available and required spinning reserves, where the red line represents the required reserve and the green line represents the actual reserve.

Analyzing these curves, it is clear that this constraint affects the optimal solution because the algorithm cannot turn off the ICEs earlier; otherwise, the available spinning reserve would become too low.

In the next subsection, the proposed approach is compared with a standard optimization method across several scenarios.

6.3. Comparison with Standard Approach

After analyzing the optimization process using the proposed problem codification and SNO method, the method was compared with a standard approach. Using the standard approach, an optimization process is considered here in which all the free problem variables (power production of dispatchable generators and SoC trajectory of the BESS) are directly managed by the optimizer. This standard approach is referred to in the following as full optimization.

The comparison was conducted across 28 different scenarios, optimizing with SNO and performing 50 independent trials. Using diverse scenarios demonstrates the approach's robustness to variations in loads and RES production.

Figure 13 shows a comparison in terms of the consumption of the final solution. Each bar represents the average value, and the 5th and the 95th percentiles are represented by the black whiskers. The green bars represent the analyzed methodology, and the orange ones represent full optimization. The figure shows the superiority of the proposed approach in terms of both the average value and standard deviation across all the analyzed scenarios in absolute terms. In this way, the actual advantages in terms of fuel consumption when the microgrid is managed with the optimization method proposed are highlighted. In addition, it is possible to compare the two methods also in relative terms. On average, compared to a standard scheduling approach, our optimization methodology leads to an improvement of 11.87%. Therefore, the proposed approach shows a significant reduction in fuel consumption in both absolute and relative terms.

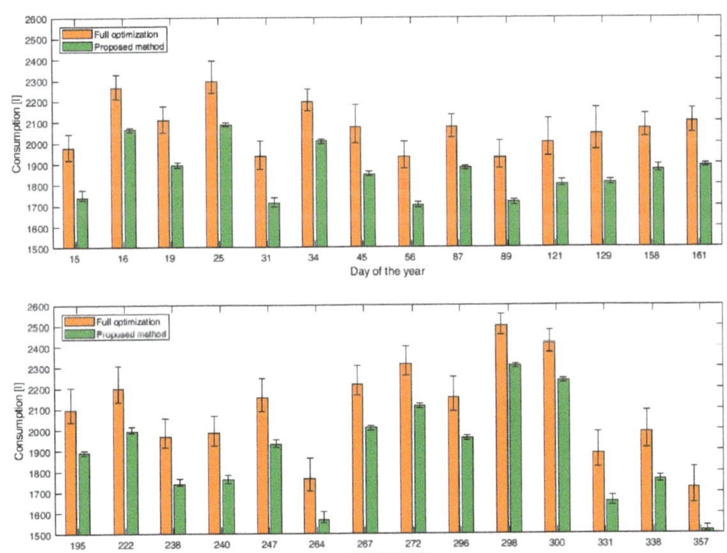

Figure 13. Comparison between full optimization and the proposed approach across 28 different scenarios. The bars represent the average values of 50 independent trials, and the black whiskers represent the 5th and the 95th percentiles.

The advantages of the proposed method can also be analyzed using the convergence curves for one of the sample scenarios, as shown in Figure 14, which highlights that both the final value and reliability improved with the proposed approach. Indeed, in this case, the average value obtained by the 50 independent trials is lower, and the confidence band is thinner.

In order to identify where the improvement in performance is introduced and the proposed approach's reliability lies, the cost contributions were analyzed for all the scenarios, as shown in Figure 15. In particular, the figure shows how the variation in the final fuel consumption was obtained; all the bars are positive because the consumptions obtained by the proposed method were lower than those found by the standard approach. Each bar is divided into two parts: blue represents the amount of fuel saved by reducing the power production of the diesel engine, and brown represents the fuel saved using a better selection of the working point of the generators at the same value of power output.

These results show that the improvement is, on average, driven by lower energy production. This is due to the reduced number of optimization variables and the introduction of the energy balance in the variable decodification process. The remaining improvement is due to better specific consumption, which is due to the separate minimization problem used to define the diesel generators' power breakdown.

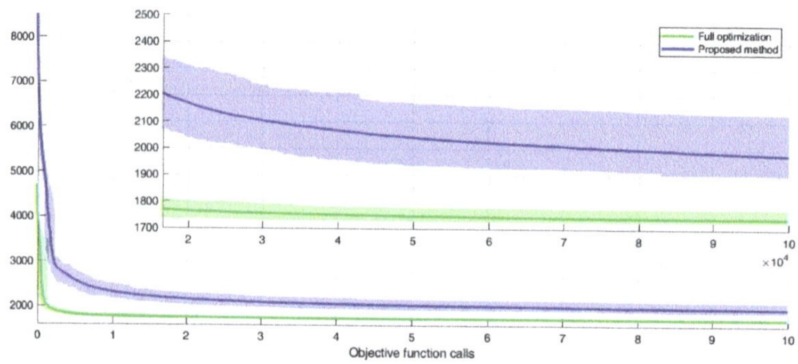

Figure 14. Comparison of the convergence curves of the proposed method (in green) and full optimization (in purple). The thick lines represent the average value of 50 independent trials, and the shaded area represents the 98% confidence band.

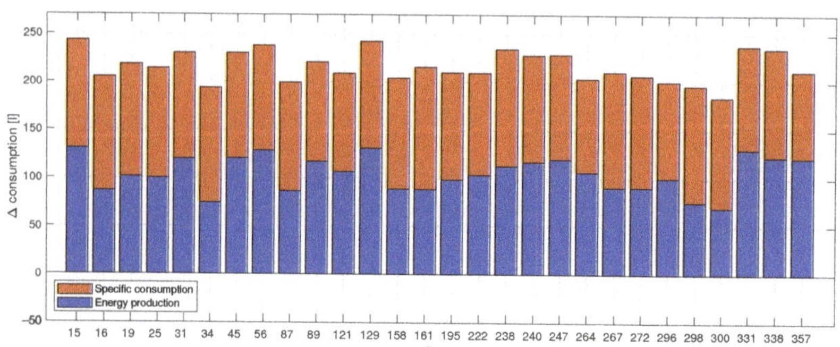

Figure 15. Components of the variation in fuel consumption between the proposed method and full optimization: blue represents the fuel saved by the proposed approach due to lower energy consumption, while brown represents the fuel saved due to better specific consumption.

6.4. Improved Storage Energy Management

The implemented Energy Management System improves storage management by addressing two key factors: reducing energy losses in the BESS and minimizing the number of charge-discharge cycles.

The energy losses are primarily determined by the BESS's efficiency, and the optimization algorithm identifies a solution that operates at the maximum efficiency point. Figure 16 shows this improvement, with the data normalized to the total daily energy demand.

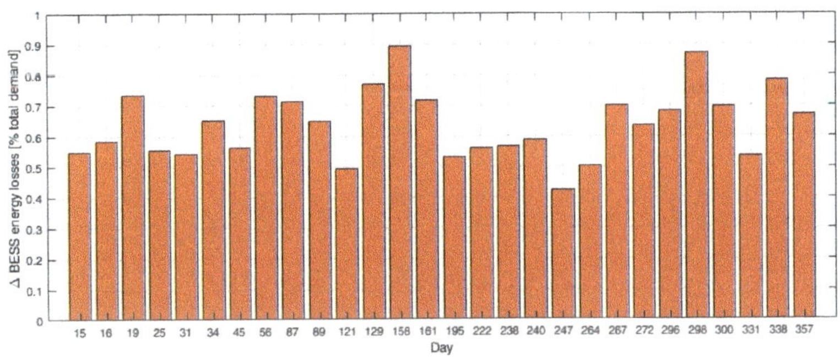

Figure 16. Difference in energy losses between basic optimization and the proposed method. The data are normalized with respect to the total daily energy demand.

The degradation of the battery can be analyzed by computing the number of cycles performed by the battery on each of the tested days. Figure 17 shows the comparison between the basic optimization and the proposed method on the number of battery cycles. The results take into account 50 independent trials. The bars represent the average values, and the whiskers the 5th and the 95th percentiles.

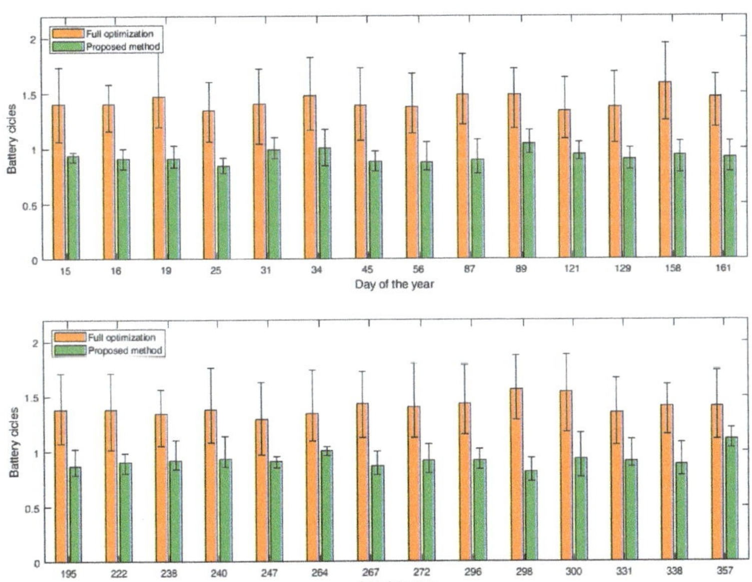

Figure 17. Comparison of the number of battery cycles between full optimization and the proposed approach across 28 different scenarios. The bars represent the average values of 50 independent trials, and the black whiskers represent the 5th and the 95th percentiles.

The results show that the proposed method is able to reduce the average number of cycles per day from 1.42 to 0.92, corresponding to an improvement of 35%.

7. Conclusions

In this paper, evolutionary algorithms were applied to optimize scheduling decisions in the Energy Management System of a microgrid. The proposed approach is capable of finding the optimal trade-off between the degrees of freedom of the optimization algorithm and the heuristic information introduced in the optimization process. The proposed approach was tested with different evolutionary algorithms, and Social Network Optimization achieved the best performance. This procedure was compared with a traditional approach in which the optimization algorithm can tune all of the free design variables. The results show that the heuristic information introduced is very important for the quality of the final result. In addition, the proposed method enhances battery management by accounting for both energy losses and battery degradation.

The results obtained in this work can serve as a benchmark for the application of reinforcement learning. These approaches are particularly promising, as they can be implemented in real time and are inherently more robust to the uncertainties associated with loads and weather forecasting. However, their deployment requires a dedicated training phase for each new microgrid configuration, which can be a limiting factor. Furthermore, RL methods are highly data-intensive, necessitating the availability of large and representative datasets to achieve satisfactory performance.

Author Contributions: Conceptualization, A.N., A.G. and S.L.; Data curation, A.N., L.F.B. and S.T.; Formal analysis, R.Z. and S.L.; Investigation, A.N. and S.T.; Methodology, A.G. and R.Z.; Project administration, S.L. and F.G.; Resources, A.N. and F.G.; Software, S.T. and L.F.B.; Supervision, A.G. and S.L.; Validation, L.F.B. and R.Z.; Visualization, A.N. and F.G.; Writing—review and editing, A.N. and S.T. All authors have read and agreed to the published version of the manuscript.

Funding: This study was carried out within MOST (The National Center for Sustainable Mobility) and received funding from the European Union's Next-GenerationEU (PIANO NAZIONALE DI RIPRESA E RESILIENZA (PNRR)—MISSIONE 4 COMPONENTE 2, INVESTIMENTO 1.4—D.D. 1033 17/06/2022, CN00000023). This manuscript reflects only the authors' views and opinions; neither the European Union nor the European Commission can be considered responsible for them. This work was supported by the Italian Ministry for Education, University, and Research under the project PRIN2022 "BERENICE - Battery energy management systems for renewable and citizen energy communities", identified by reference number P2022R3L83, which is part of the National Recovery and Resilience Plan (PNRR).

Data Availability Statement: The data presented in this study are available on request from the corresponding author due to privacy concerns about electricity consumption data.

Conflicts of Interest: The authors declare no conflicts of interest.

Abbreviations

The following abbreviations are used in this manuscript:

BBO	Biogeography-Based Optimization
BESS	Battery Energy Storage System
CI	Computational Intelligence
DE	Differential Evolution
EA	Evolutionary Algorithm
EMS	Energy Management System

EV	Electric Vehicle
FIS	Fuzzy Inference System
GA	Genetic Algorithm
MILP	Mixed-Integer Linear Programming
PSO	Particle Swarm Optimization
RES	Renewable Energy Sources
SoC	State of Charge
SNO	Social Network Optimization
List of Symbols	
$P_{G_i}(t)$	Power production of the i-th dispatchable generator
$P_{MPP}(t)$	Maximum power produced by RES
$P_{RES}(t)$	Effective power input from RES
$P_{curt}(t)$	Curtailed power from RES
$P_d(t)$	Total load power absorption
$P_{slack}(t)$	Load slack power
$P_L(t)$	Effective load power absorption
P_{ch}	BESS charging power
P_{dh}	BESS discharging power
$P_{dump}(t)$	Dumped power
C_N	BESS nominal capacity
$P_{ch,max}$ / $P_{dh,max}$	BESS maximum charging/discharging power
ΔSoC^+ / ΔSoC^-	BESS SoC variation in charge/discharge
Δt	Time discretization interval
η^+ / η^-	BESS charging/discharging efficiency
$F_i(t)$	Fuel consumption of the i-th dispatchable generator
$\eta_{G_i}(P_{G_i})$	Efficiency of the i-th dispatchable generator at P_{G_i} power
$P_{G_i,min}$ / $P_{G_i,max}$	Minimum/maximum power of the i-th dispatchable generator
SoC_{min}	BESS minimum state of charge
SoC_{max}	BESS maximum state of charge
$SR_{av}(t)$	Available spinning reserve
$z_i(t)$	On/off status of the i-th dispatchable generator

References

1. Zidane, T.E.K.; Ab Muis, Z.; Ho, W.S.; Zahraoui, Y.; Aziz, A.S.; Su, C.L.; Mekhilef, S.; Campana, P.E. Power systems and microgrids resilience enhancement strategies: A review. *Renew. Sustain. Energy Rev.* **2025**, *207*, 114953. [CrossRef]
2. Gholami, M.; Muyeen, S.; Mousavi, S.A. Optimal sizing of battery energy storage systems and reliability analysis under diverse regulatory frameworks in microgrids. *Energy Strategy Rev.* **2024**, *51*, 101255. [CrossRef]
3. Bhatia, A.A.; Das, D. Demand response strategy for microgrid energy management integrating electric vehicles, battery energy storage system, and distributed generators considering uncertainties. *Sustain. Energy Grids Netw.* **2025**, *41*, 101594. [CrossRef]
4. Ali, L.; Azim, M.I.; Peters, J.; Bhandari, V.; Menon, A.; Tiwari, V.; Green, J.; Muyeen, S. Application of a community battery-integrated microgrid in a blockchain-based local energy market accommodating P2P trading. *IEEE Access* **2023**, *11*, 29635–29649. [CrossRef]
5. Micallef, A.; Guerrero, J.M.; Vasquez, J.C. New horizons for microgrids: From rural electrification to space applications. *Energies* **2023**, *16*, 1966. [CrossRef]
6. Eyimaya, S.E.; Altin, N. Review of Energy Management Systems in Microgrids. *Appl. Sci.* **2024**, *14*, 1249. [CrossRef]
7. Thirunavukkarasu, G.S.; Seyedmahmoudian, M.; Jamei, E.; Horan, B.; Mekhilef, S.; Stojcevski, A. Role of optimization techniques in microgrid energy management systems—A review. *Energy Strategy Rev.* **2022**, *43*, 100899. [CrossRef]
8. Alghamdi, A.S. Microgrid energy management and scheduling utilizing energy storage and exchange incorporating improved gradient-based optimizer. *J. Energy Storage* **2024**, *97*, 112775. [CrossRef]
9. Bilal, M.; Algethami, A.A.; Hameed, S. Review of computational intelligence approaches for microgrid energy management. *IEEE Access* **2024**, *12*, 123294–123321. [CrossRef]
10. Nespoli, A.; Ogliari, E.; Pretto, S.; Gavazzeni, M.; Vigani, S.; Paccanelli, F. Electrical Load Forecast by Means of LSTM: The Impact of Data Quality. *Forecasting* **2021**, *3*, 91–101. [CrossRef]

11. Niccolai, A.; Dolara, A.; Ogliari, E. Hybrid PV Power Forecasting Methods: A Comparison of Different Approaches. *Energies* **2021**, *14*, 451. [CrossRef]
12. Gharehveran, S.S.; Shirini, K.; Khavar, S.C.; Mousavi, S.H.; Abdolahi, A. Deep learning-based demand response for short-term operation of renewable-based microgrids. *J. Supercomput.* **2024**, *80*, 26002–26035. [CrossRef]
13. Jumani, T.A.; Mustafa, M.W.; Alghamdi, A.S.; Rasid, M.M.; Alamgir, A.; Awan, A.B. Swarm intelligence-based optimization techniques for dynamic response and power quality enhancement of AC microgrids: A comprehensive review. *IEEE Access* **2020**, *8*, 75986–76001. [CrossRef]
14. Ahmad, S.; Shafiullah, M.; Ahmed, C.B.; Alowaifeer, M. A review of microgrid energy management and control strategies. *IEEE Access* **2023**, *11*, 21729–21757. [CrossRef]
15. Papazoglou, G.; Biskas, P. Review and comparison of genetic algorithm and particle swarm optimization in the optimal power flow problem. *Energies* **2023**, *16*, 1152. [CrossRef]
16. Mohapatra, P.; Roy, S.; Das, K.N.; Dutta, S.; Raju, M.S.S. A review of evolutionary algorithms in solving large scale benchmark optimisation problems. *Int. J. Math. Oper. Res.* **2022**, *21*, 104–126. [CrossRef]
17. Piotrowski, A.P.; Napiorkowski, J.J.; Piotrowska, A.E. Particle swarm optimization or differential evolution—A comparison. *Eng. Appl. Artif. Intell.* **2023**, *121*, 106008. [CrossRef]
18. Pradhan, P.C.; Sahu, R.K.; Panda, S. Comparative performance analysis of hybrid differential evolution and pattern search technique for frequency control of the electric power system. *J. Electr. Syst. Inf. Technol.* **2021**, *8*, 1–21. [CrossRef]
19. Moretti, L.; Meraldi, L.; Niccolai, A.; Manzolini, G.; Leva, S. An Innovative Tunable Rule-Based Strategy for the Predictive Management of Hybrid Microgrids. *Electronics* **2021**, *10*, 1162. [CrossRef]
20. Zhang, Q.; Wei, L.; Yang, B. Research on Improved BBO Algorithm and Its Application in Optimal Scheduling of Micro-Grid. *Mathematics* **2022**, *10*, 2998. [CrossRef]
21. Ahmethodžić, L.; Musić, M.; Huseinbegović, S. Microgrid energy management: Classification, review and challenges. *CSEE J. Power Energy Syst.* **2022**, *9*, 1425–1438.
22. Sun, C.; Joos, G.; Ali, S.Q.; Paquin, J.N.; Rangel, C.M.; Al Jajeh, F.; Novickij, I.; Bouffard, F. Design and real-time implementation of a centralized microgrid control system with rule-based dispatch and seamless transition function. *IEEE Trans. Ind. Appl.* **2020**, *56*, 3168–3177. [CrossRef]
23. Kassab, F.A.; Celik, B.; Locment, F.; Sechilariu, M.; Liaquat, S.; Hansen, T.M. Optimal sizing and energy management of a microgrid: A joint MILP approach for minimization of energy cost and carbon emission. *Renew. Energy* **2024**, *224*, 120186. [CrossRef]
24. Moretti, L.; Polimeni, S.; Meraldi, L.; Raboni, P.; Leva, S.; Manzolini, G. Assessing the impact of a two-layer predictive dispatch algorithm on design and operation of off-grid hybrid microgrids. *Renew. Energy* **2019**, *143*, 1439–1453. [CrossRef]
25. Zhu, X.; Ruan, G.; Geng, H.; Liu, H.; Bai, M.; Peng, C. Multi-objective sizing optimization method of microgrid considering cost and carbon emissions. *IEEE Trans. Ind. Appl.* **2024**, *60*, 5565–5576. [CrossRef]
26. Capillo, A.; De Santis, E.; Mascioli, F.M.F.; Rizzi, A. An Online Hierarchical Energy Management System for Energy Communities, Complying with the Current Technical Legislation Framework. *arXiv* **2024**, arXiv:2402.01688.
27. Leonori, S.; Paschero, M.; Mascioli, F.M.F.; Rizzi, A. Optimization strategies for Microgrid energy management systems by Genetic Algorithms. *Appl. Soft Comput.* **2020**, *86*, 105903. [CrossRef]
28. Leonori, S.; Martino, A.; Mascioli, F.M.F.; Rizzi, A. Microgrid energy management systems design by computational intelligence techniques. *Appl. Energy* **2020**, *277*, 115524. [CrossRef]
29. Gholami, K.; Dehnavi, E. A modified particle swarm optimization algorithm for scheduling renewable generation in a micro-grid under load uncertainty. *Appl. Soft Comput.* **2019**, *78*, 496–514. [CrossRef]
30. Bandopadhyay, J.; Roy, P.K. Application of hybrid multi-objective moth flame optimization technique for optimal performance of hybrid micro-grid system. *Appl. Soft Comput.* **2020**, *95*, 106487. [CrossRef]
31. Mazzola, S.; Vergara, C.; Astolfi, M.; Li, V.; Perez-Arriaga, I.; Macchi, E. Assessing the value of forecast-based dispatch in the operation of off-grid rural microgrids. *Renew. Energy* **2017**, *108*, 116–125. [CrossRef]
32. Moretti, L.; Astolfi, M.; Vergara, C.; Macchi, E.; Pérez-Arriaga, J.I.; Manzolini, G. A design and dispatch optimization algorithm based on mixed integer linear programming for rural electrification. *Appl. Energy* **2019**, *233*, 1104–1121. [CrossRef]
33. Farakhor, A.; Wu, D.; Wang, Y.; Fang, H. A novel modular, reconfigurable battery energy storage system: Design, control, and experimentation. *IEEE Trans. Transp. Electrif.* **2022**, *9*, 2878–2890. [CrossRef]
34. Zheng, N.; Qin, X.; Wu, D.; Murtaugh, G.; Xu, B. Energy storage state-of-charge market model. *IEEE Trans. Energy Mark. Policy Regul.* **2023**, *1*, 11–22. [CrossRef]
35. Niccolai, A.; Grimaccia, F.; Mussetta, M.; Gandelli, A.; Zich, R. Social network optimization for WSN routing: Analysis on problem codification techniques. *Mathematics* **2020**, *8*, 583. [CrossRef]
36. Li, M.; Zhang, D.; Lu, S.; Tang, X.; Phung, T. Differential evolution-based overcurrent protection for DC microgrids. *Energies* **2021**, *14*, 5026. [CrossRef]

37. de Oliveira-Assis, L.; García-Trivino, P.; Soares-Ramos, E.P.; Sarrias-Mena, R.; García-Vázquez, C.A.; Ugalde-Loo, C.E.; Fernández-Ramírez, L.M. Optimal energy management system using biogeography based optimization for grid-connected MVDC microgrid with photovoltaic, hydrogen system, electric vehicles and Z-source converters. *Energy Convers. Manag.* **2021**, *248*, 114808. [CrossRef]
38. Firdouse, F.; Surender Reddy, M. A hybrid energy storage system using GA and PSO for an islanded microgrid applications. *Energy Storage* **2023**, *5*, e460. [CrossRef]

Disclaimer/Publisher's Note: The statements, opinions and data contained in all publications are solely those of the individual author(s) and contributor(s) and not of MDPI and/or the editor(s). MDPI and/or the editor(s) disclaim responsibility for any injury to people or property resulting from any ideas, methods, instructions or products referred to in the content.

energies

MDPI

Article

Comprehensive Control Strategy for Hybrid Energy Storage System Participating in Grid Primary Frequency Regulation

Haorui Jiang [1], Kuihua Han [1,*], Weiyu Bao [2] and Yahui Li [1,*]

1 Shandong Engineering Research Center for High-Efficiency Energy Storage and Hydrogen Energy Utilization, Shandong University, Jinan 250061, China; 202334564@mail.sdu.edu.cn

2 School of Electrical Engineering, Shandong University, Jinan 250061, China; baoweiyu@sdu.edu.cn

* Correspondence: hankh@163.com (K.H.); liyahui@sdu.edu.cn (Y.L.)

Abstract: The increasing integration of renewable energy sources has posed significant challenges to grid frequency stability. To maximize the advantages of energy storage in primary frequency regulation, this paper proposes a comprehensive control strategy for a hybrid energy storage system (HESS) based on supercapacitor battery. Firstly, considering the characteristics of the HESS and different control strategies, the battery responds to virtual droop control to reduce frequency deviation, while the supercapacitor responds to inertia control to suppress frequency drops and facilitate frequency recovery. Simultaneously, a reasonable dynamic dead zone is configured to prevent frequent actions of the battery and thermal unit while allowing flexible adjustments according to the load condition. Thirdly, an algebraic S-curve-based adaptive droop coefficient incorporating SOC is proposed, while the inertia coefficient additionally considers load type, enhancing adaptability. Furthermore, to better maintain the battery's SOC, an improved adaptive recovery strategy within the battery dead zone is proposed, considering both SOC recovery requirements and system frequency deviation constraints. Finally, a simulation validation was conducted in MATLAB/Simulink. Compared to the conventional strategy, the proposed control strategy reduces the frequency drop rate by 17.43% under step disturbance. Under compound disturbances, the RMS of frequency deviation decreases by 13.34%, and the RMS of battery SOC decreases by 68.61%. The economic benefit of this strategy is 3.212 times that of the single energy storage scheme. The results indicate that the proposed strategy effectively alleviates sudden frequency disturbances, suppresses frequency fluctuations, and reduces battery output while maintaining the SOC of both the supercapacitor and the battery, thereby extending the battery lifespan and improving economic performance.

Keywords: adaptive control strategy; hybrid energy storage system; primary frequency regulation; virtual droop control; virtual inertia control

1. Introduction

Under the depletion of fossil energy sources and the development of clean energy sources, the global energy structure is undergoing profound changes, and wind and solar energy are increasing steadily. The rapid development of these clean energy sources provides new impetus for achieving sustainable energy development and environmental protection [1,2].

The high percentage of renewable energy integration results in low inertia characteristics in the system, making it challenging to maintain frequency stability [3,4]. Traditional thermal power units (TPUs) are constrained by ramp rate limitations, and they face challenges such as slow response, poor regulation accuracy, and reverse regulation, and the frequency regulation standards of the grid are hard to satisfy [5]. Energy storage systems (ESSs) are widely used in primary frequency regulation (PFR) applications and have better features such as faster response, bi-directional regulation, and precise tracking capabilities [6]. For example, a universal mathematical model for ESSs was developed in reference [7], laying the groundwork for future studies. Equating ESS to a first-order inertia model simplifies the model without affecting experimental results was also demonstrated in reference [8].

When the system frequency experiences sudden changes, increasing system inertia and adjusting power output based on frequency variation can both reduce frequency deviation (FD) [9]. Virtual droop control (VDC) and virtual inertia control (VIC) were separately introduced in references [10,11]. These strategies enable ESS to simulate the regulation and dynamic characteristics of TPUs, providing a basis for subsequent research. On the one hand, the VDC adjusts the active power output of ESS based on the deviation between the grid frequency and the rated frequency, effectively altering the steady-state value of the FD [12,13]. It was applied to electric vehicles for the purpose of grid PFR, thereby enhancing the voltage stability of DC distribution networks [14]. On the other hand, VIC uses the rate of change in the grid FD as the control signal to simulate the inertia response of conventional generators, effectively mitigating grid frequency degradation. In reference [15], it was proposed that battery energy storage systems (BESSs) could adopt both VIC and VDC methods in PFR. The VIC can respond promptly to large disturbances, thereby preventing the exacerbation of frequency deterioration, but it impedes frequency recovery during the restoration phase [16]. Therefore, virtual negative inertia control (VNIC) was proposed in reference [17]; it is opposite in sign to VIC and can facilitate frequency recovery during the restoration phase. Then, a switching method between VIC and VDC for PFR was employed in references [18,19], where the switching timing is determined through sensitivity analysis. This method combines the advantages of VIC and VDC, but the ESS output experiences a significant jump when switching, which easily causes frequency fluctuations. To smooth the output during switching, the direct switching of output modes was modified in reference [20] by distributing the output ratio between VDC and VIC using a distribution coefficient, thereby ensuring a smoother output transition. However, this approach typically depends on complex functions, making it challenging to implement in engineering practices.

Extensive research has focused on the control coefficients for ESS charging and discharging. The "fixed-K method" was employed in reference [21], which involves a constant VDC coefficient to control the BESS output. In reference [22], the method is improved, which is called the "variable-K method", and it is categorized into ordinary, conservative, aggressive, and hybrid types. The VDC coefficient considers the SOC of the BESS, as shown in reference [23], enhancing frequency regulation effectiveness while preventing overcharging and over-discharging of the BESS. Most studies determined control coefficients based on the SOC of energy storage systems, employing functions such as hyperbolic tangent [24], piecewise [25], quartic [26], or linear [27,28]. However, these methods were basically adapted to changes in SOC and did not consider variations in control coefficients under different load conditions. Meanwhile, excessively complex curves increase the computational burden during actual operation.

As the requirements for grid frequency stability continue to increase, a single ESS has frequently proved to be inadequate in meeting these demands. ESS can be categorized into power-type storage and energy-type storage [29]. Energy-type electrochemical BESS has the advantages of high energy density and relatively low cost [30]. However, frequent grid frequency fluctuations lead to frequent BESS cycling, significantly shortening the lifespan of the BESS. Power-type energy storage systems, such as the supercapacitor energy storage system (SESS), are widely used. It has extremely high charge/discharge rates and a long cycle life, enabling rapid response, but its low energy density prevents long-duration operation. Currently, no ESS can meet the requirements for both high power density and high energy density [31]. Therefore, HESS can overcome the limitations of a single storage system, providing complementary advantages and more effectively maintaining grid stability.

Existing research on HESS coordination faces the problem of simply combining two types of storage and the lack of flexibility in control strategies. Flywheel ESS and BESS were directly connected in parallel in reference [32] to allow each to leverage the advantages of the respective energy storage types. Reference [33] employed a sliding average filtering method to allocate wind power, where the flywheel absorbs high-frequency power and the lithium battery energy storage system handles low-frequency power. Reference [34] selected a lithium battery and a supercapacitor for ESS frequency regulation, each using VDC, achieving better frequency regulation than a single ESS. However, the selection of control strategies is somewhat limited. References [16,35] simply applied control methods previously used for the BESS to both the BESS and SESS, combining them without addressing the coordination between different energy storage types or solving the problems inherent in the control strategies of individual ESS.

In summary, the deficiencies in the current ESS control strategies for PFR are as follows:

(1) Current control strategies do not fully leverage the characteristics of ESSs and fundamental control methods;
(2) Control coefficients for ESS output are fixed or only based on SOC without considering output variations under different operating conditions, while a more reasonable output curve is lacking.
(3) The potential for SOC recovery within the PFR dead zone is not fully utilized.

To resolve these problems, a comprehensive control strategy for a battery-supercapacitor HESS participating in PFR is presented. The power-type SESS responds to the inertia control with VIC during frequency deterioration and VNIC during frequency recovery, making use of the high power of the SESS and inertia control to rapidly suppress frequency deterioration and facilitate frequency recovery. The large-capacity BESS responds to the VDC signal, continuously supplying power to reduce FD. A dynamically adjustable dead zone based on load conditions is configured, and the SESS is prioritized for power output through reasonable interval settings. To better maintain the SOC while regulating frequency, an algebraic S-curve-based adaptive droop coefficient incorporating SOC is proposed, with the VIC coefficient further considering the load types to enhance adaptability. To address the issue of SOC of BESS deterioration, adaptive recovery of BESS SOC is applied within the dead zone of the BESS, considering both SOC recovery requirements and system frequency deviation constraints. Finally, the comprehensive control strategy was verified through simulations under different disturbances, demonstrating its effectiveness across various metrics. It also reduces the charge/discharge cycles of BESS and enhances its cycle life, yielding better economic performance.

2. PFR Control Model Based on HESS

This section develops a dynamic model for PFR of a regional grid, incorporating HESS. Then, the advantages and disadvantages of VDC, VIC, and VNIC are theoretically analyzed. Finally, two typical types of loads are analyzed and differentiated.

2.1. Primary Frequency Regulation Model

Based on the regional equivalent method, the models of each component within the region are converted into transfer function form. The resulting PFR model is shown in Figure 1. The parameters and units of the model are provided in Table 1.

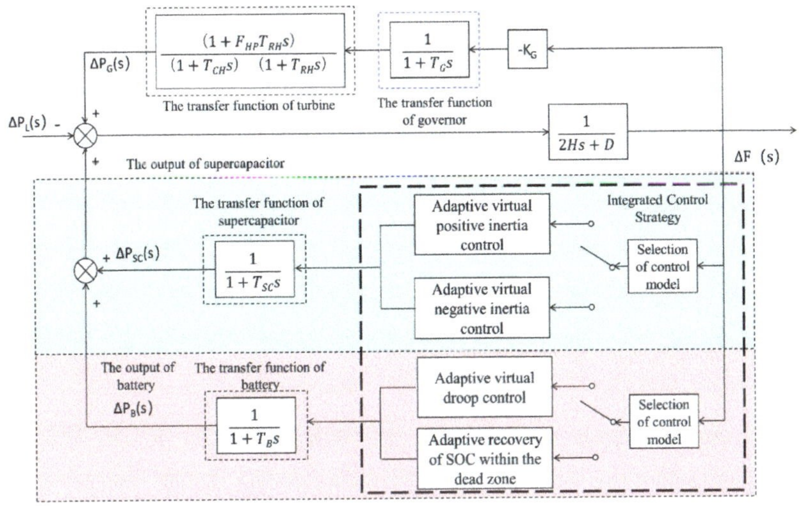

Figure 1. Dynamic model of PFR in regional power grid with HESS.

Table 1. Definition of parameters in the model.

Parameters	Definition	Unit
$\Delta P_G(s)$	PFR output of the TPU	MW
$\Delta P_{SC}(s)$	SESS PFR power out	MW
$\Delta P_B(s)$	BESS PFR power out	MW
$\Delta P_L(s)$	Load disturbance	MW
K_G	PFR factor for TPU	-
H	Inertia TC of the power grid	-
D	Load damping coefficient	-
F_{HP}	Reheater gain	-
T_G	Governor TC of the TPU	s
T_{RH}	Reheater TC	s
T_{CH}	Turbine TC	s
T_{SC}	Inertia TC of SESS	s
T_B	Inertia TC of BESS	s

Where the TC presents the time constant.

2.2. Three Control Strategies

Following frequency fluctuations in the power grid, the frequency regulation signal passes through the prime mover and governor of the TPU, leading to a change in the turbine steam inlet flow, which in turn modifies the mechanical power output [36]. ESS can mimic the control of traditional TPUs to offset FD, as shown in Equation (1),

$$\begin{gathered}\Delta f(s) = \frac{\Delta P_G(s)+\Delta P_E(s)-\Delta P_L(s)}{2Hs+D} \\ \Delta P_G(s) = -K_G \Delta f(s) G_g(s) \\ G_g(s) = \frac{1+F_{HP}T_{RH}s}{(1+T_{CH}s)(1+T_{RH}s)} \\ \Delta P_E(s) = \Delta P_{K_E}(s) + \Delta P_{M_E}(s)\end{gathered}, \quad (1)$$

where $G_g(s)$ represents the transfer function of the TPU, and $\Delta P_E(s)$ represents the output of the ESS. The output of the VDC and VIC is $\Delta P_{M_E}(s)$ and $\Delta P_{K_E}(s)$, given by (2). The principle of VNIC resembles VIC, with only the output direction reversed, and therefore does not require separate analysis.

$$\begin{aligned} \Delta P_{M_E}(s) &= -M_E s \Delta f(s) G_E(s) \\ \Delta P_{K_E}(s) &= -K_E \Delta f(s) G_E(s) \ , \\ s\Delta f(s) &= \mathrm{d}\Delta f(s)/\mathrm{d}t \end{aligned} \tag{2}$$

where $s\Delta f(s)$ represents the rate of change in the FD, M_E represents the VIC coefficient, and K_E represents the VDC coefficient. Substituting Equation (2) into Equation (1) yields the following:

$$\begin{aligned} \Delta f_{M_E}(s) &= \frac{-\Delta P_L(s)}{2Hs + D - K_G G_g(s) - M_E s G_E(s)} \\ \Delta f_{K_E}(s) &= \frac{-\Delta P_L(s)}{2Hs + D - K_G G_g(s) - K_E G_E(s)} \end{aligned} . \tag{3}$$

Analyzing the rate of change in frequency deviation $\Delta o_0(s)$ and the quasi-steady-state frequency deviation Δf_{qs} for VDC yields the following:

$$\begin{aligned} \Delta o_0 &= \lim_{s\to\infty} s \cdot [s \cdot \Delta f_{K_E}(s)] = \frac{-\Delta p_L}{2H} \\ \Delta f_{qs} &= \lim_{s\to 0} s \cdot \Delta f_{K_E}(s) = \frac{-\Delta p_L}{D + K_G + K_E} \end{aligned} . \tag{4}$$

As shown in Equation (4), VDC effectively reduces the steady-state FD but has no impact on the rate of change in FD.

Analyzing the rate of change in frequency deviation $\Delta o_0(s)$ and the quasi-steady-state frequency deviation Δf_{qs} for VIC yields the following:

$$\begin{aligned} \Delta o_0 &= \lim_{s\to\infty} s \cdot [s \cdot \Delta f_{M_E}(s)] = \frac{-\Delta p_L}{2H + M_E} \\ \Delta f_{qs} &= \lim_{s\to 0} s \cdot \Delta f_{M_E}(s) = \frac{-\Delta p_L}{D + K_G} \end{aligned} . \tag{5}$$

As shown in Equation (5), VIC does not impact the final steady-state FD but does influence the rate of change in FD. Positive inertia control can effectively reduce the rate of change in FD, preventing its further deterioration, but it impedes the recovery of the FD. During the frequency recovery phase, VNIC facilitates the restoration of frequency. The proper coordination of VIC and VNIC can better utilize the ESS throughout the entire frequency regulation period.

2.3. Analysis of Load Disturbance Types

Substituting Equation (2) into Equation (1) yields the following:

$$(2H + M_E)\frac{\mathrm{d}\Delta f}{\mathrm{d}t} + (D + K_E)\Delta f = \Delta P_G - \Delta P_L. \tag{6}$$

At the moment the disturbance is introduced into the system at time t_0, the TPU has not yet responded, and the frequency has not changed. Therefore, Equation (6) can be expressed as Equation (7), which takes the form of Equation (8).

$$(2H + M_E)\frac{\mathrm{d}\Delta f}{\mathrm{d}t}\,|_{t=t_0} = -\Delta P_L\,|_{t=t_0} . \tag{7}$$

$$\frac{\mathrm{d}\Delta f}{\mathrm{d}t}\,|_{t=t_0} = \frac{\Delta P_L\,|_{t=t_0}}{(2H + M_E)} . \tag{8}$$

From this, it can be concluded that the rate of change in system FD is closely related to the type of disturbance introduced into the system.

There are two typical load disturbance scenarios in power grids: one is the sudden event disturbance (which is also called step disturbance), and the other is load fluctuation (which is also called continuous disturbance) [12]. Sudden event disturbances are usually associated with the startup or shutdown of large-capacity motors or load shedding due to grid faults. These disturbances in the power grid are characterized by sudden changes, leading to rapid and significant frequency variations. Load fluctuations are typically caused by the variability of renewable energy, random load variations in the power system, and uncertainties in grid operation. These disturbances are sustained over time, unlike step disturbances that are sudden and significant.

Based on the characteristics of the two load types, the disturbance type can be determined using the rate of change in FD, as illustrated by Equation (9),

$$\Delta o(t) = \frac{f_i - f_{i-1}}{\Delta T}, \tag{9}$$

where f_i and f_{i-1} denote the frequency at the i-th and the $i-1$-th sampling point. ΔT is the sampling time interval.

The time-varying curves of $\Delta o(t)$ under two typical disturbances are shown in Figure 2.

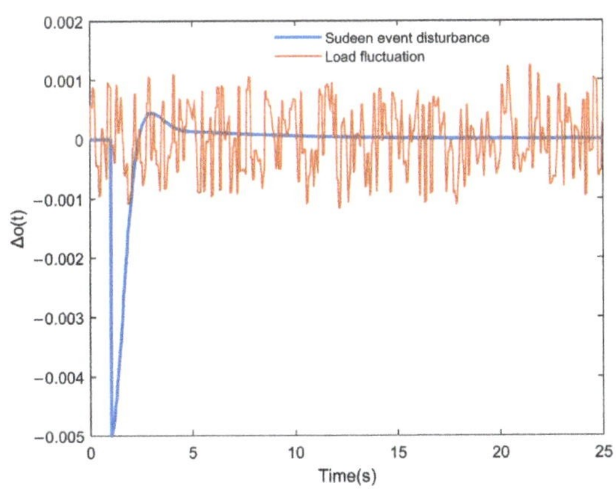

Figure 2. Curves of the rate of change in FD under different disturbances.

Under a sudden disturbance, $\Delta o(t)$ rapidly increases to its peak and then quickly returns to zero. Under load fluctuations, $\Delta o(t)$ fluctuates slightly around zero. Since the $\Delta o(t)$ caused by sudden events varies across different power grids, the threshold value for determining a sudden disturbance also differs accordingly. In practical applications, the critical empirical value of o_{ref} is decided in accordance with the actual conditions of the regional power grid. When $\Delta o(t) > \Delta o_{\text{ref}}$, it indicates a sudden event disturbance.

Reference [37] theoretically demonstrated that the rate of change in FD at the initial moment under step disturbance can be calculated using Equation (10),

$$\frac{\mathrm{d}\Delta f}{\mathrm{d}t}\Big|_{t=t_0} = \frac{f_s}{(2H + M_E)R}, \tag{10}$$

where Δf_s is the FD at steady state after the step disturbance.

GB/T 15945-2008 [38] states that under normal operating conditions of the power system, the absolute value of FD should be within 0.2 Hz, so the steady-state FD should be less than 0.2 Hz. Substituting f_s = 0.2 Hz and the relevant parameters of this model (see Section 5.1) into Equation (10), it is determined that when the disturbance causes the rate of change of the FD to reach 0.18 Hz/s, the system's steady-state deviation will exceed the safe limit. Reference [39] mentions that during the UK blackout incident, the rate of change of the FD at the onset of the disturbance was 0.135 Hz/s. Therefore, for safety considerations, the value of o_{ref} should not exceed 0.135 Hz/s.

A quasi-steady-state coefficient δ is introduced to determine whether the frequency reaches a quasi-steady state after a sudden event disturbance. The expression is shown in Equation (11), where $\delta < \delta_{\text{ref}}$ indicates that the system frequency has entered a quasi-steady state. δ_{ref} is set to 0.001 [40].

$$\delta = \frac{f_i - f_{i-1}}{f_i}. \tag{11}$$

3. Control Strategies for ESS Participation in PFR

3.1. Coordinated Control Based on Supercapacitor-Battery HESS

As analyzed before, VDC can provide a continuous output over the entire frequency regulation cycle to eliminate the FD. During the grid frequency drop phase, VIC is required to respond quickly and mitigate frequency changes. During the recovery phase, VNIC is required to support the restoration process. Each of the three control methods has its own characteristics and application scenarios, and achieving optimal frequency regulation requires their coordination throughout the entire regulation cycle. Reference [41] applied VIC and VNIC along with VDC to a single ESS, leveraging the advantages of these three control methods. Reference [42] employed both BESS and SESS for frequency regulation but only used VDC. Reference [16] utilized HESS and coordinated the three control strategies within a single storage device, achieving effective frequency regulation, but did not adequately account for the specific requirements of VIC and VDC on the storage characteristics.

BESS is the most widely used ESS, known for its high energy density, low power density, relatively short cycle life, and low cost. In contrast, SESS exhibits high power density and long cycle life, but its standalone application in PFR is limited due to its smaller capacity and higher cost. Based on the characteristics of the two types of ESS and the three control strategies, SESS, with its high power and rapid response characteristics, is well-suited to respond to inertial signals. BESS, with its high energy density and low cost, is more appropriate for VDC and can operate effectively throughout the entire frequency regulation cycle.

Therefore, to optimally match different ESSs and control methods, this paper proposes a coordination control strategy for HESS participation in PFR. When the FD exceeds the frequency regulation dead zone, the BESS operates with VDC to eliminate the FD. During the frequency deterioration phase, where $[\mathrm{d}\Delta f/\mathrm{d}t]\cdot\Delta f > 0$, the SESS uses VIC to counteract the frequency decline. In the frequency recovery phase, where $[\mathrm{d}\Delta f/\mathrm{d}t]\cdot\Delta f < 0$, the SESS employs VNIC to promote frequency recovery.

Figure 3 presents the Bode plots analyzing the magnitude–frequency characteristics under four scenarios: no ESS participation, BESS-only frequency regulation, SESS-only frequency regulation, and HESS frequency regulation.

It can be observed that the BESS, governed by VDC, effectively suppresses frequency deviation in the low-frequency range. The SESS, utilizing VIC, mitigates the impact of high-frequency power fluctuations on system frequency. The HESS strategy integrates the advantages of both control approaches. Across the full frequency spectrum, it yields lower and flatter response curves compared to single strategies. In the low-frequency range, its minimum magnitude ensures optimal steady-state accuracy; in the mid-frequency to high-frequency bands, it smooths out gain peaks of the single strategies, enhancing robustness to multi-frequency disturbances; and the steeper high-frequency attenuation aids in noise suppression.

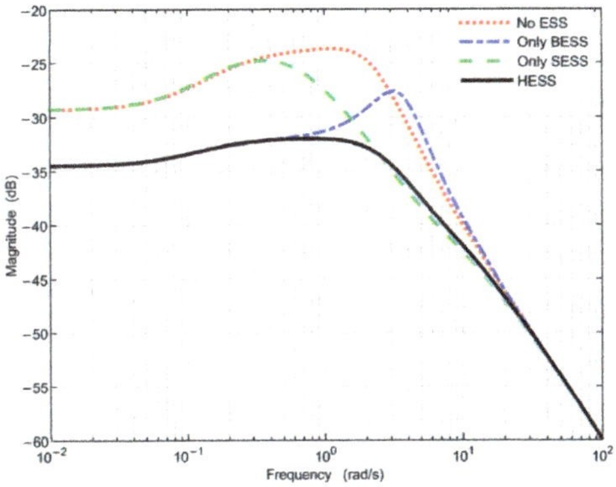

Figure 3. Amplitude–frequency characteristic of different strategies.

The gain margins (GM) of the four strategies, calculated using MATLAB (R2022b), are 24.16 dB, 30.05 dB, 24.99 dB, and 32.22 dB, respectively. The HESS approach yields the highest GM, implying superior robustness against gain fluctuations. For the linearized closed-loop systems under all four strategies, all poles are strictly located in the left half of the complex plane, confirming the stability of each strategy.

Additionally, to further leverage the fast response and long cycle life of SESS, its dead zone $\Delta f_{SC}{}^{d}$ is set smaller than that of BESS, ensuring that SESS responds to frequency variations first. This prioritization enables frequency restoration before reaching the dead zone of the BESS ($\Delta f_{B}{}^{d}$), significantly reducing the number of BESS cycles and mitigating its lifespan degradation due to cycling. Meanwhile, the $\Delta f_{B}{}^{d}$ is set smaller than that of the TPU to prevent frequent start-stop operations, which could lead to mechanical wear.

To account for the varying demands for dead zones under different load types and time periods, a dynamic dead zone is introduced for the ESS in this study. The expression of the dynamic dead zone is shown in Equation (12),

$$\text{Dead Zone}(t, \Delta f) = k_1(\Delta f)k_2(t)f_G^d, \tag{12}$$

where k_1 is the load adjustment coefficient that varies dynamically with the load, and k_2 is the time adjustment coefficient that differs between peak and off-peak hours, $f_G{}^{d}$ is the dead zone for the TPU.

The construction rule of k_1 is as follows: under high-frequency, low-amplitude fluctuations, a larger dead zone is adopted to reduce ineffective cycling and equipment wear; under low-frequency and large-amplitude disturbances, a smaller dead zone is employed to ensure rapid frequency regulation response. The expression of k_1 is provided in Equation (13),

$$k_1(f) = \begin{cases} k_{1\min} + \frac{\Delta f_{th}}{|\Delta f|}(k_{1\max} - k_{1\min}) & |\Delta f| \geq \Delta f_{th} \\ k_{1\max} & |\Delta f| < \Delta f_{th} \end{cases}, \tag{13}$$

where $k_{1\max}$ is the maximum value of the load adjustment coefficient (set to 0.6 for the SESS and 0.8 for the BESS), and $k_{1\min}$ is the minimum value (0.55 for the SESS and 0.75 for the BESS). Δf_{th} is the reference value of FD; a deviation greater than Δf_{th} indicates significant load disturbance. Based on multiple simulation validations, Δf_{th} is set to 0.1 Hz in this study.

The construction principle of k_2 is that during electricity peak hours (generally from 17:00 to 22:00), both residential and commercial areas experience high and fluctuating loads, causing many small-magnitude but frequent short-term disturbances. Appropriately enlarging the dead zone during these times helps ignore small but frequent disturbances within a safe margin, thereby reducing unnecessary switching of storage devices.

The expression of k_2 is given in Equation (14),

$$k_2(t) = \begin{cases} 1.1\ 17:00 < t < 22:00 \\ 1 \quad \text{elsetimeperiods} \end{cases}. \tag{14}$$

3.2. Adaptive Droop Coefficient Based on SOC Feedback

Traditional virtual droop control often uses fixed droop coefficients, known as the "fixed-K method". This method maintains a constant power output throughout the entire frequency regulation cycle, achieving good results in the early stages of frequency fluctuation. However, it does not take into account the SOC of the ESS, which can lead to overcharging and over-discharging, significantly affecting the lifetime of the ESS. To ensure effective frequency regulation performance while also considering the SOC of the ESS, using an adaptive virtual droop coefficient that incorporates the SOC is a good choice. When the SOC is within an optimal range, the ESS prioritizes frequency regulation needs, charging and discharging at higher power levels. When the SOC is excessively low or high, the charging/discharging power is reduced to balance frequency regulation performance and the SOC. When the SOC exceeds the threshold, charging and discharging are stopped to prevent overcharging or over-discharging, which could pose a risk to the ESS.

To meet these requirements, most studies employ the logistic function to constrain the power output [43–45]. Reference [46] points out that the design of output constraints for ESS should consider the feasibility and engineering convenience in subsequent hardware implementation. Reference [47] states that complex mathematical operations such as exponential functions can cause significant computational burdens in embedded systems, whereas algebraic S-curves offer superior computational efficiency. Reference [48] demonstrates that algebraic S-curves meet the requirements for high-speed and high-precision robot applications. Reference [49] notes that in automatic machinery trajectory planning, algebraic S-curves facilitate a trade-off between smooth motion and computational cost.

The virtual droop coefficient based on the algebraic S-curve is shown in Equations (15) and (16).

When $\Delta f > 0$,

$$K_c = \begin{cases} K_{\max} & SOC \le SOC_{high} \\ K_{\max} \times [1 - (3(\frac{SOC - SOC_{high}}{SOC_{\max} - SOC_{high}})^2 - 2(\frac{SOC - SOC_{high}}{SOC_{\max} - SOC_{high}})^3)] & SOC_{high} < SOC < SOC_{\max} \\ 0 & SOC \ge SOC_{\max} \end{cases}. \tag{15}$$

When $\Delta f < 0$,

$$K_d = \begin{cases} 0 & SOC \le SOC_{\min} \\ K_{\max} \times [(3(\frac{SOC - SOC_{\min}}{SOC_{low} - SOC_{\min}})^2 - 2(\frac{SOC - SOC_{\min}}{SOC_{low} - SOC_{\min}})^3)] & SOC_{\min} < SOC < SOC_{low} \\ K_{\max} & SOC \ge SOC_{low} \end{cases}, \tag{16}$$

where K_c and K_d represent the virtual droop coefficient during the charging and discharging of the ESS, respectively. SOC indicates the measured SOC of the ESS. $SOC_{\min}$, SOC_{low}, SOC_{high}, and $SOC_{\max}$ represent the minimum, low, high, and maximum SOC levels, respectively, varying with different types of ESS.

The virtual droop coefficients for SESS charging and discharging are K_{SCc} and K_{SCd}, respectively, with $SOC_{\max} = 0.9$, $SOC_{\text{high}} = 0.55$, $SOC_{\text{low}} = 0.45$, and $SOC_{\min} = 0.1$. It should be noted that the SESS in the proposed control strategy does not employ VDC. However, the concept of the virtual droop control coefficient is used to help construct its virtual inertia control coefficient. For BESS charging and discharging, the virtual droop coefficients are K_{Bc} and K_{Bd}, with SOC_{max} = 0.8, SOC_{high} = 0.55, SOC_{low} = 0.45, and SOC_{min} = 0.2.

Taking the SESS as an example, the relationship between the virtual droop coefficient and SOC is shown in Figure 4.

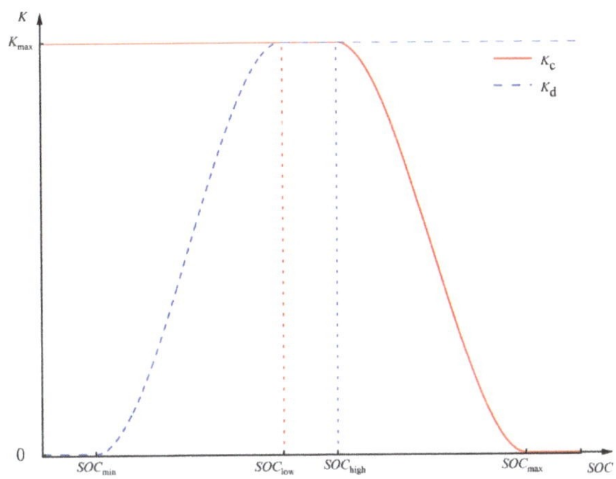

Figure 4. Relationship between VDC coefficient and SOC.

Figure 5 compares charge–discharge curves over the SOC interval [0.1, 0.45] and [0.55, 0.9] using algebraic S-curves, linear functions, logistic functions, and hyperbolic tangent functions (the values of the hyperbolic tangent function are approximately zero at the endpoints). It can be observed that linear functions exhibit poor adaptability in output curves, whereas algebraic S-curves, logistic functions, and hyperbolic tangent functions can effectively describe the transition from rapid response to steady state, achieving good adaptive performance. In practical engineering applications, the algebraic S-curve can reduce the computational burden and enhance real-time responsiveness without sacrificing control accuracy.

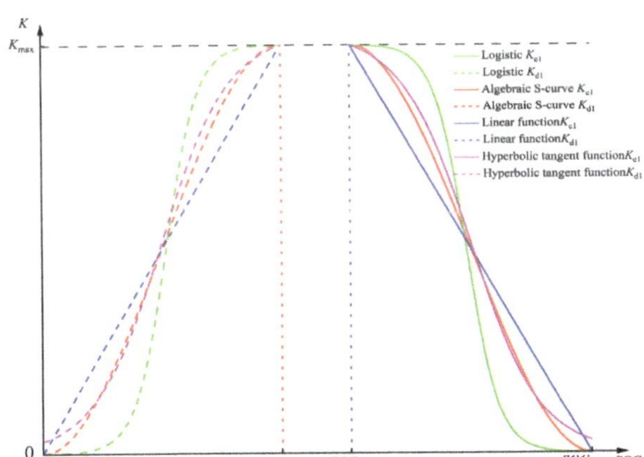

Figure 5. Comparison of different VDC curves.

3.3. Adaptive Inertia Coefficient Based on Disturbance Type and SOC

The virtual inertia control coefficient M_{EP} is expressed by Equation (17),

$$M_{EP} = \begin{cases} \alpha\beta\gamma K_{SCc}, \Delta f \geq 0 \\ \alpha\beta\gamma K_{SCd}, \Delta f < 0 \end{cases}, \tag{17}$$

where α is the output adjustment factor. Since the magnitudes of the FD and the rate of change in FD differ significantly, a proper value of α should be set to ensure comparable outputs of VIC and VDC. In this study, α is set to 0.5 [50–52].

β is the load adjustment factor, which varies with the type of load. As analyzed in Section 2.3, under load fluctuation, the rate of change in FD is relatively small; a moderate increase in the inertia coefficient can assist in mitigating frequency fluctuations. At this time, β is set to 1. During a sudden event disturbance, the system FD and its rate of change are large. Using a high inertia coefficient during such instances may result in excessive power injection/absorption, thereby leading to a new frequency fluctuation. At this time, β should take a smaller value of 0.5 [53]. When $\delta < \delta_{ref}$, it implies the FD is stabilizing, permitting β to return to a higher value of 1.

γ is the direction adjustment factor.

Reference [54] presents sensitivity curves for different values of M_E and finds that the sensitivity changes from negative to positive and then tends toward zero. Further sensitivity analysis on M_E demonstrates that VIC is effective only before the FD reaches its peak. After that, it suppresses frequency recovery; the suppression becomes more significant with a larger M_E.

Therefore, γ is introduced to regulate the direction of power output. When $d\Delta f/dt \cdot \Delta f > 0$, indicating that the FD and its rate of change have the same direction, the system is in a frequency deterioration phase, γ takes a positive value, indicating that VIC is applied to mitigate frequency deterioration. When $d\Delta f/dt \cdot \Delta f \leq 0$, the FD and its rate of change are opposite, the system is in a frequency recovery phase ($d\Delta f/dt \cdot \Delta f \leq 0$), and γ becomes negative, signifying the use of VNIC to promote frequency recovery.

An extreme disturbance (0.1 p.u.) was introduced into the system using only virtual inertia control; the maximum FD reached 0.15 Hz, staying within the safe limit, demonstrating the robustness of the VID coefficient in this model.

3.4. SOC Recovery Within the BESS Frequency Regulation Dead Zone

Previous research has shown that the SOC of the BESS often struggles to remain stable during large-scale system disturbances [17]. Therefore, in addition to the adaptive BESS output during the frequency regulation phase, an adaptive recovery mechanism is introduced within the BESS's dead zone. Within the dead zone, where grid frequency remains within acceptable limits, the recovery demand coefficient is calculated according to the SOC. To prevent excessive charging or discharging power during recovery, which could push the frequency outside the dead zone again, a recovery constraint coefficient is derived considering the grid's tolerance capacity. The SOC recovery coefficient is determined based on both the recovery demand coefficient and the recovery constraint coefficient. SESS, on the other hand, can effectively maintain its SOC under various operating conditions and therefore does not require an additional SOC recovery mechanism.

3.4.1. Recovery Demand Coefficient Determination Based on SOC

Taking the charging recovery demand coefficient K_{c1} as an example, when $SOC < SOC_{min}$, the SOC is excessively low, which may negatively impact the lifespan of the BESS. Under such conditions, the BESS charges at maximum power, and K_{c1} reaches its maximum value.

When $SOC_{min} < SOC < SOC_{low}$, K_{c1} remains relatively high but decreases as SOC increases. In the range $SOC_{low} < SOC < SOC_{high}$, the SOC is considered to be within a safe range, and no recovery is required. When SOC > SOC_{high}, the SOC is already high, and further charging could worsen SOC conditions. Therefore, for $SOC > SOC_{low}$, K_{c1} is set to 0. The specific expressions for the charging and discharging recovery demand coefficients K_{c1} and K_{d1} are given in Equations (18) and (19).

$$K_{c1} = \begin{cases} K_{max} & SOC \le SOC_{min} \\ K_{max} \times [1 - (3(\frac{SOC - SOC_{min}}{SOC_{low} - SOC_{min}})^2 - 2(\frac{SOC - SOC_{min}}{SOC_{low} - SOC_{min}})^3)] & SOC_{min} < SOC < SOC_{low} \\ 0 & SOC \ge SOC_{low} \end{cases} . \tag{18}$$

$$K_{d1} = \begin{cases} 0 & SOC \le SOC_{high} \\ K_{max} \times [(3(\frac{SOC - SOC_{high}}{SOC_{max} - SOC_{high}})^2 - 2(\frac{SOC - SOC_{high}}{SOC_{max} - SOC_{high}})^3)] & SOC_{high} < SOC < SOC_{max} \\ K_{max} & SOC \ge SOC_{max} \end{cases} . \tag{19}$$

The recovery demand coefficients for charge/discharge are illustrated in Figure 6.

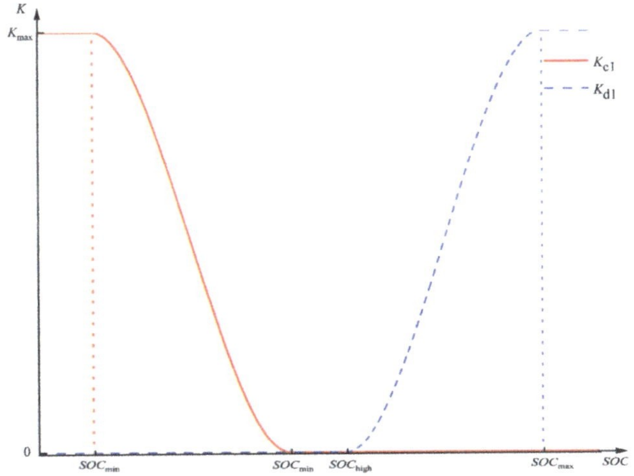

Figure 6. The coefficient of demand for charge/discharge recovery.

3.4.2. Recovery Constraint Coefficient Based on Δf

The FD is classified into six states: $-\Delta f_{d}$, $-\Delta f_{high}$, $-\Delta f_{low}$, Δf_{low}, Δf_{high} and Δf_{d}, where Δf_{low}, Δf_{high} and Δf_{d} correspond to the lower value, higher value, and upper dead zone limit of the FD, respectively.

Taking the charging recovery constraint coefficient K_{c2} as an example, the principle is as follows: when Δf is less than $-\Delta f_{high}$, the grid frequency is in a dangerous state with a risk of exiting the dead zone, and the discharge of the BESS is required to adjust the FD; hence, K_{c2} is set to 0. When Δf lies between $-\Delta f_{high}$ and $-\Delta f_{low}$, the grid frequency is relatively safe, and K_{c2} increases as the absolute value of the FD decreases. When Δf lies between $-\Delta f_{low}$ and Δf_{low}, the frequency is within a safe range. When Δf exceeds Δf_{low}, the grid requires the BESS charging. Therefore, for $\Delta f > -\Delta f_{low}$, K_{c2} reaches its maximum value. The charge/discharge recovery constraint coefficients K_{c2} and K_{d2} are constructed using a cosine function, with their specific expressions shown in Equations (20) and (21).

$$K_{c2} = \begin{cases} 0 & -f_d < f < -f_{\text{high}} \\ K_{\max} \times \frac{1-\cos\left(\pi \frac{f-(-f_{\text{high}})}{-f_{\text{low}}-(-f_{\text{high}})}\right)}{2} & -f_{\text{high}} \le f \le -f_{\text{low}} \\ K_{\max} & -f_{\text{low}} < f < f_d \end{cases}. \tag{20}$$

$$K_{d2} = \begin{cases} K_{\max} & -f_d < f < f_{\text{low}} \\ K_{\max} \times \frac{1+\cos\left(\pi \frac{f-f_{low}}{f_{\text{high}}-f_{\text{low}}}\right)}{2} & f_{\text{low}} \le f \le f_{\text{high}} \\ 0 & f_{\text{high}} < f < f_d \end{cases}. \tag{21}$$

The recovery constraint coefficients for charge/discharge are illustrated in Figure 7.

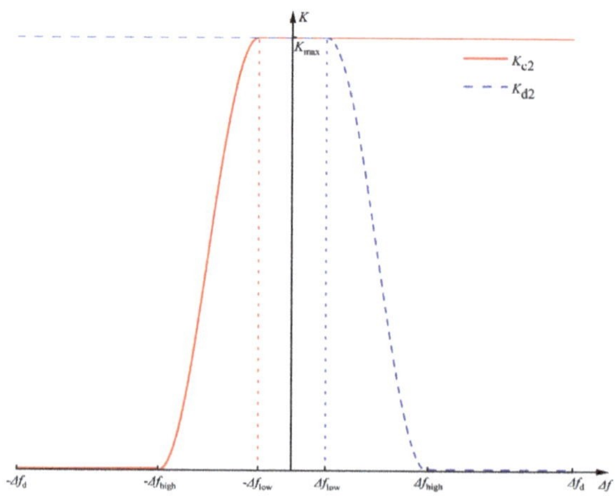

Figure 7. The coefficient of constraint for charge/discharge recovery.

3.4.3. Determination of the Recovery Coefficient

To balance the SOC recovery demand of the BESS with the FD constraint of the grid, most studies selected the smaller value between the recovery demand coefficient and the recovery constraint coefficient as the SOC recovery coefficient of the BESS [55–57]. However, this approach leads to abrupt changes in BESS output during recovery, causing secondary disturbances to the system. Moreover, selecting the minimum value imposes excessive restrictions on recovery power, making the strategy overly conservative. To achieve smoother power output during battery recovery, a demand-constraint dynamically coupled recovery coefficient is adopted, as expressed in Equation (22).

$$K_{\text{requ}} = \begin{cases} a \cdot K_{c1} + (1-a)K_{c2} \; S \leqslant SOC_{low} \\ 0 \qquad SOC_{low} < S < SOC_{high} \\ a \cdot K_{d1} + (1-a)K_{d2} \; S \geqslant SOC_{high} \end{cases}. \tag{22}$$

The value of a is determined by Equation (23),

$$a = \frac{1}{1+e^{-k_1(SOC-0.5)}} \times \left(1 - \frac{|\Delta f|}{\Delta f_d}\right)^{k_2}, \tag{23}$$

where k_1 and k_2 are tuning coefficients used to adjust the magnitude of a.

When the SOC deviates significantly from 0.5 while the frequency remains within a safe range, a approaches 1, prioritizing the SOC recovery demand. When the SOC is close to 0.5 but the FD is about to exceed the dead zone, a tends toward 0, prioritizing frequency constraints. In intermediate states, a varies between 0 and 1, with its weight determined based on the deviations of SOC and Δf.

The calculation of the BESS recovery output P_{requ} within the dead zone is expressed in Equation (24),

$$\Delta P_{\text{requ}} = -K_{\text{requ}}|\Delta f|. \tag{24}$$

4. Comprehensive Control Strategy and Evaluation Metrics

4.1. Comprehensive Control Strategy

This study puts forward a comprehensive control method for HESS to participate in PFR by considering the characteristics of SESS and BESS, flexibly adopting the three control strategies, and integrating adaptive control rules based on SOC and load type. The specific control strategy is illustrated in Figure 8.

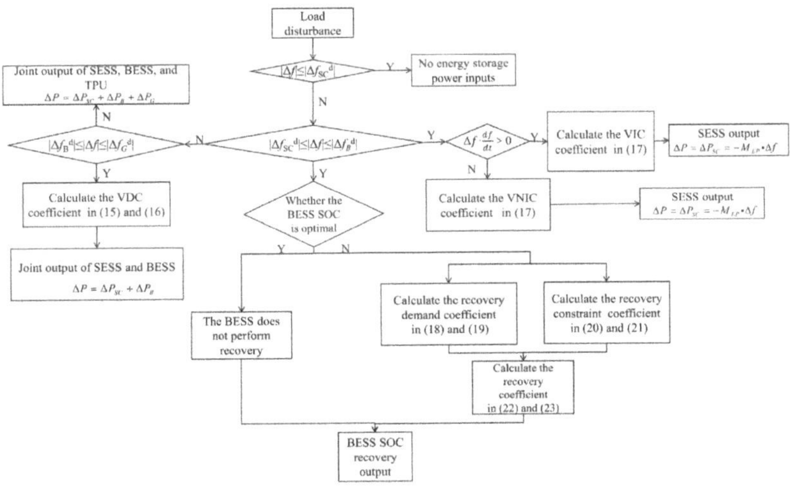

Figure 8. Comprehensive control method of HESS participating in PFR of power grid.

Taking the case where $\Delta f > 0$, indicating that the BESS needs to absorb power for frequency regulation, yields the following:

(1) When $\Delta f < \Delta f_{\text{SC}}{}^{\text{d}}$, the FD lies within a safe range, so no frequency regulation is required.

(2) When $\Delta f_{\text{SC}}{}^{\text{d}} \ll \Delta f \ll \Delta f_{\text{B}}{}^{\text{d}}$, the SESS is employed for frequency regulation, while the BESS decides whether to restore its SOC based on its current status.

① The participation of the SESS in grid frequency regulation can be categorized as follows:

(a) When $\text{d}\Delta f/\text{d}t > 0$: At this stage, the FD and its rate of change are in the same direction, indicating a worsening frequency condition. The SESS determines the VIC coefficient based on (17) and applies adaptive VIC to mitigate frequency deterioration.

(b) When $\text{d}\Delta f/\text{d}t \ll 0$: Here, the FD and its rate of change are in opposite directions, indicating a frequency recovery phase. The SESS determines the VNIC coefficient based on (17) and employs adaptive VNIC to speed up frequency recovery.

② The recovery of BESS SOC within the dead zone is governed by the following principles:

(a) When the SOC of the BESS is in a satisfactory condition, the recovery of SOC is deemed unnecessary.

(b) When the SOC of the BESS is suboptimal, the recovery demand coefficient is determined based on (18) and (19), while the recovery constraint coefficient is calculated based on (20) and (21). Considering both the recovery demand and recovery constraint, the recovery coefficient is determined according to Equations (22) and (23), which is then used to calculate the SOC recovery power output of the BESS.

(3) For $\Delta f_B{}^d \ll \Delta f \ll \Delta f_G{}^d$, the SESS and BESS collaborate for frequency regulation. The BESS participates in grid frequency regulation using VDC, with the VDC coefficient calculated according to (15) and (16). While the SESS continues to apply positive or negative inertia control based on the product of the FD and its rate of change.

(4) For $\Delta f > \Delta f_G{}^d$, the FD is large, and the HESS collaborates with the TPU to generate power.

From the above process, it can be seen that the proposed comprehensive control strategy divides the output timing of HESS and TPUs through dead zones, determines the output mode and direction based on FD and its rate of change, and adapts the droop and inertia coefficients according to SOC and disturbance type.

4.2. Evaluation Metrics

For step disturbance loads, the commonly used evaluation metrics are as follows: Δf_m and t_m represent the maximum FD and the corresponding time after the disturbance, while Δf_s and t_s represent the steady-state FD and the corresponding time. v_m denotes the rate of frequency decline, and $v_m = |\Delta f_m| / t_m$. A lower Δf_m implies a less significant frequency drop, a smaller v_m reflects a slower rate of frequency decline, while a smaller Δf_s signifies better frequency recovery. Consequently, the smaller the values of these three parameters, the more effective the frequency regulation method.

For continuous disturbance signals, the root mean square (RMS) of FD, termed f_{index}, and the RMS of SOC deviation, termed SOC_{index}, are utilized as evaluation indicators, with the specific forms provided in Equations (25) and (26),

$$f_{index} = \sqrt{\frac{1}{n}\sum_{i=1}^{n} \Delta f_i^2}, \tag{25}$$

$$SOC_{index} = \sqrt{\frac{1}{n}\sum_{i=1}^{n} (SOC_i - SOC_0)^2}, \tag{26}$$

where Δf_i represents the FD at the i-th sampling point, SOC_i represents the SOC at the i-th sampling point, SOC_0 is the set reference value, which is taken as 0.5, and N signifies the total number of sampling points. f_{index} and SOC_{index} reflect the degree of dispersion of frequency and SOC, respectively. It is evident that the smaller the values of these indices, the smaller the deviation of frequency and SOC from the reference standard, indicating better frequency regulation and SOC maintenance performance of the system.

To evaluate the economic performance, an economic assessment model was established based on the life cycle cost theory and estimated benefits of ESS participation in PFR.

The cost present value C_{LCC} is expressed as shown in Equation (27),

$$C_{LCC} = C_{inv} + C_{O\&M} + C_{scr} + C_\beta + C_\alpha, \tag{27}$$

where C_{inv} is the life cycle investment cost of ESS, $C_{O\&M}$ is the operation and maintenance cost, C_{scr} is the disposal cost, C_β is the penalty cost due to deficiencies, and C_α is the cost of curtailed energy. The detailed formulation can be found in reference [54].

The net revenue present value N_{RES} is calculated as in Equation (28),

$$N_{REG} = R_{REG} \times E_{REG}, \tag{28}$$

where R_{REG} is the market compensation for PFR and E_{REG} is the regulation energy.

The net present benefit P_{NET} is given in Equation (29),

$$P_{NET} = N_{RES} - C_{LCC}. \tag{29}$$

A higher P_{NET} indicates better economic performance of the proposed control strategy.

5. Simulation Verification

5.1. Simulation Model

To evaluate the effectiveness of the proposed control strategy, a PFR simulation model for a regional power grid, as illustrated in Figure 1, was constructed in MATLAB/Simulink (R2022b). The thermal power unit has a rated capacity of 1000 MW, with the SESS configured at 25 MW/0.5 MW·h and the BESS capacity set at 10 MW/1 MW·h. To better capture the variations in the SOC of the ESSs, the ESSs were configured with relatively small capacities. The system reference frequency is 50 Hz, and the primary PFR dead zone for the TPU was set at ±0.033 Hz. The remaining parameters were converted into per-unit values based on the rated frequency and rated capacity of the unit. Specifically, the F_{HP} value was set to 0.3; T_G, T_{RH}, and T_{CH} were set to 0.1, 10, and 0.3 s, respectively; H was set to 5; D was set to 4; and T_{SC} and T_B were set to 0.2 and 0.3, respectively.

To explore the appropriate value for o_{ref}, a segment of actual grid load data shown in Figure 9 was introduced to the HESS, and the recognition effectiveness of the step disturbance signal was tested at three values of 0.05 Hz/s, 0.1 Hz/s, and 0.135 Hz/s.

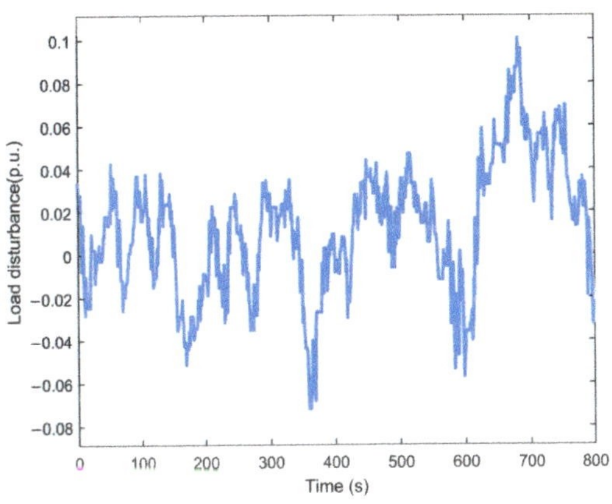

Figure 9. Variation curve of realistic load disturbance curve for 800 s.

Under the same load disturbance, the system with o_{ref} set to 0.05 Hz/s identified 8262 sudden disturbance events, resulting in a very high false detection rate, which was completely unacceptable. The system with o_{ref} set to 0.135 Hz/s failed to recognize any sudden disturbance events, resulting in a high miss detection rate, which could not meet the requirements. The system with o_{ref} set to 0.1 Hz/s recognized five sudden disturbance events, which was more in line with the actual grid situation.

To investigate the impact of different o_{ref} values on system stability, a typical step disturbance signal was added to systems with o_{ref} values of 0.05 Hz/s, 0.1 Hz/s, and 0.135 Hz/s, and the resulting frequency deviations are shown in Figure 10. The curves for o_{ref} values of 0.05 Hz/s and 0.1 Hz/s overlap, both effectively identifying the step disturbance signal and promptly reducing the VIC coefficient, thus maintaining system stability. The system with o_{ref} set to 0.135 Hz/s failed to effectively recognize the step disturbance signal, resulting in excessive system inertia and causing frequency fluctuations.

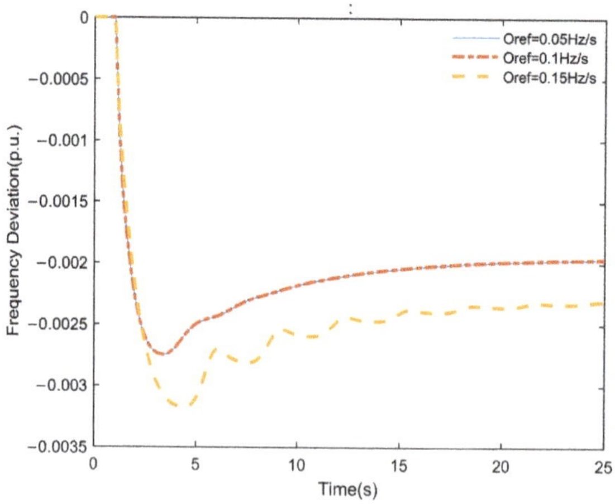

Figure 10. FD variation curves for different oref values under typical step disturbance.

In conclusion, selecting 0.1 Hz/s as the o_{ref} value for the proposed control strategy strikes a balance between recognizing step signals and maintaining system stability.

5.2. Simulation and Analysis of Step Load Disturbance

A step disturbance signal of 0.05 p.u. was employed for the model under four scenarios: the method proposed in this study ("Paper strategy"), the adaptive VDC strategy outlined in reference [58] ("Adaptive VDC strategy") (also called the conventional strategy), the fixed-K method ("Fixed-K method"), and the condition without ESS ("No ESS"). The resulting FD curves are illustrated in Figure 11, the SOC variation curves are illustrated in Figure 12a,b, and the PFR performance metrics are summarized in Table 2.

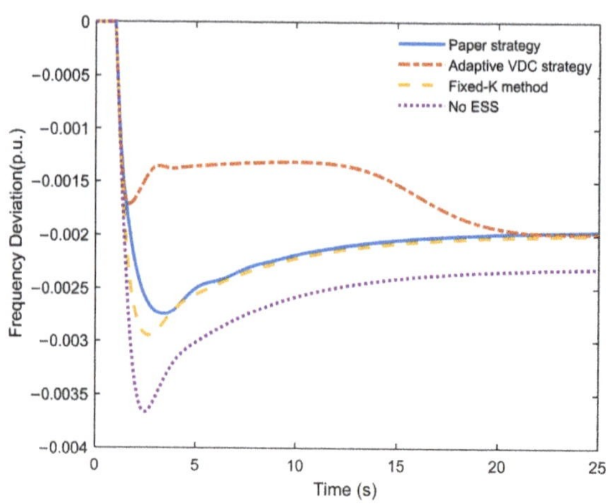

Figure 11. FD curve under 0.05 p.u. step disturbance.

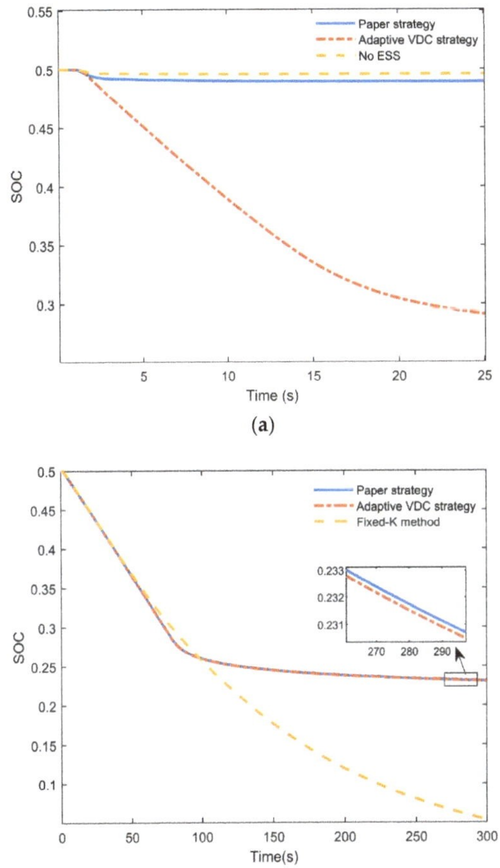

(b)

Figure 12. SOC variation curve of ESS under 0.05 p.u. step disturbance. (**a**) SOC variation curve of SESS under 0.05 p.u. step disturbance. (**b**) SOC variation curve of BESS under 0.05 p.u. step disturbance.

Table 2. PFR evaluation index under 0.05 p.u. step disturbance.

Control Policies	$\lvert \Delta f_m \rvert/10^{-3}$ p.u.	t_m	$\lvert \Delta f_s \rvert/10^{-3}$ p.u.	t_s	$v_m/10^{-3}$
Paper strategy	2.739	3.443	1.981	20.885	0.796
Adaptive VDC strategy	1.649	1.710	1.999	23.571	0.964
Fixed-K method	2.946	2.616	2.001	23.946	1.126
No ESS	3.665	2.491	2.313	24.274	1.471

As illustrated in Figure 11, the frequency of all four systems experiences a rapid decline at the initial stage of the disturbance. The system without energy storage exhibits the most significant frequency reduction, whereas the three systems with energy storage demonstrate comparatively minor declines. Although adaptive VDC achieves the smallest initial frequency drop, it experiences a secondary drop in the subsequent phase. This reveals the limitation of using VDC alone, where ESS cannot flexibly adjust its output during sudden events, leading to rapid energy depletion and an inability to continue output, causing a secondary frequency drop. The proposed control strategy achieves a smaller maximum FD compared to the other two methods. This is due to the fact that VIC can suppress the deterioration of frequency. Subsequently, the frequencies of all systems recover relatively quickly. Given that PFR is a proportional control, the FD cannot be entirely eliminated without considering the secondary frequency regulation, leaving each system's FD stabilized at a specific value. Among these, the proposed control strategy attains the smallest steady-state deviation. This demonstrates the role of using VNIC in frequency recovery. As shown in Table 2, compared to traditional TPU frequency regulation, the steady-state FD in this control strategy decreased by 25.27%, and the maximum FD decreased by 14.35%. The proposed control strategy effectively suppresses the frequency drop and facilitates frequency recovery. Compared to the other three control strategies, this strategy has the slowest frequency decline rate, with reductions of 17.43%, 29.31%, and 46.89%, respectively. Additionally, it achieves the shortest recovery time, reducing recovery time by 11.39%, 12.78%, and 13.96%, respectively. This is attributed to the combination of SESS and inertial control, which enables rapid response and adjustment.

The minimal maximum FD and absence of secondary drop indicate that this strategy effectively mitigates the impact of sudden disturbances on system frequency. Moreover, the smallest steady-state FD confirms its superior ability to mitigate the impact of sudden disturbances on grid frequency.

As illustrated in Figure 12a, under the adaptive VDC strategy, the SOC of the SESS drops significantly. This is because when only droop control is used, the ESS continues to output during large disturbances, making it difficult to maintain its SOC. Both the proposed control strategy and the fixed-K method stabilize the SOC decline of the SESS. Although the proposed strategy slightly underperforms the fixed-K method in maintaining the SOC, the difference is negligible, and the SOC remains within the safe range. Under large load disturbances, the primary task of ESS is to maintain frequency stability. Therefore, the proposed control strategy in this paper outputs more power than the fixed-K method. Sacrificing SOC within a reasonable range to achieve better frequency regulation is highly worthwhile.

As illustrated in Figure 12b, during the BESS frequency regulation process, the fixed-K method exhibits poor adaptability, causing the SOC to continuously decline beyond the acceptable range. This demonstrates the superiority of adaptive output. In contrast, both the proposed control strategy and the adaptive VDC strategy effectively stabilize the SOC decline of the BESS. The SOC maintenance capability of the proposed control strategy is nearly identical to that of the adaptive VDC strategy, slightly outperforming it, and far surpassing the fixed-K method.

Based on the preceding analysis, it can be inferred that the proposed control strategy effectively maintains the grid frequency under step disturbances and ensures the SOC stability of both the SESS and the BESS.

5.3. Simulation Analysis of Continuous Load Disturbance

To verify the PFR performance of the proposed control strategy under short-term continuous disturbance signals, a 25 s continuous disturbance signal, as shown in Figure 13, was introduced into the system. The corresponding FD is illustrated in Figure 14, while the SOC of the SESS and BESS are shown in Figure 15, and the corresponding performance metrics are listed in Table 3.

Table 3. PFR evaluation index under 25 s continuous disturbance.

Control Policies	$f_{\text{index}}/10^{-4}$ p.u.	SOC_{SCindex}	$SOC_{\text{Bindex}}/10^{-4}$
Paper strategy	2.9478	0.0049	0
Adaptive VDC strategy	4.0143	0.0019	4.3488
Fixed-K method	4.4223	0.0015	2.9782
No ESS	5.4207	-	-

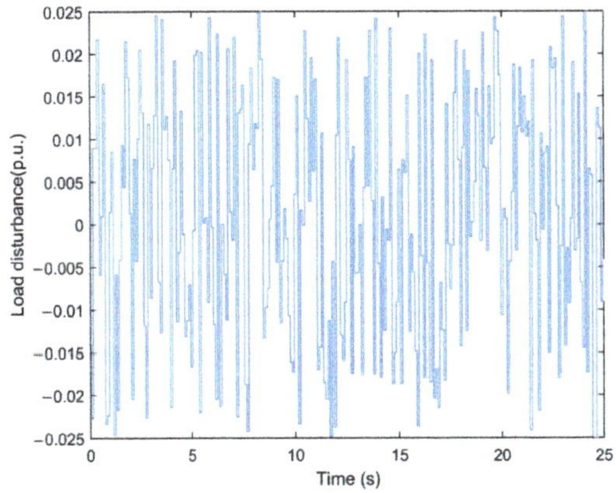

Figure 13. Variation curve of load disturbance for 25 s.

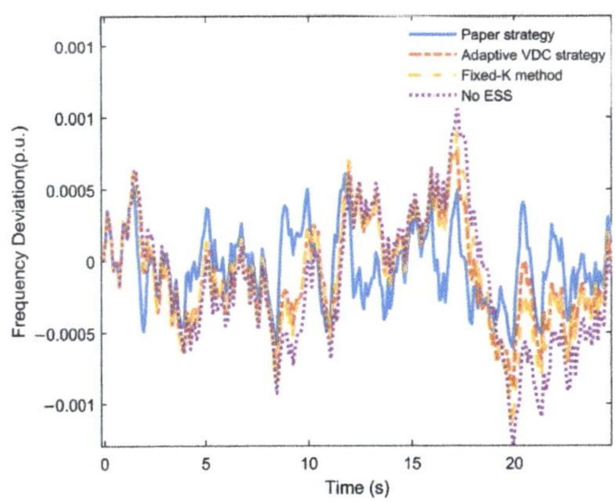

Figure 14. FD curve under 25 s continuous disturbance.

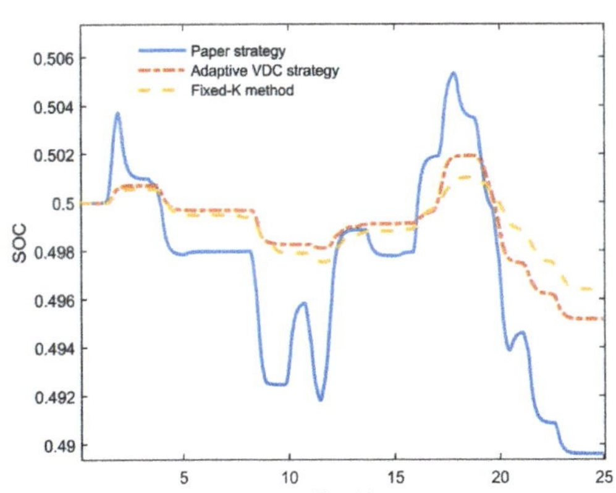

(a)

Figure 15. *Cont.*

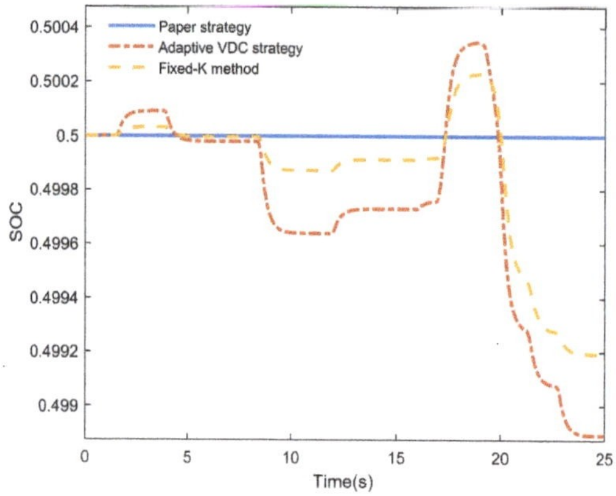

(b)

Figure 15. SOC variation curve of ESS under 25 s continuous disturbance. (**a**) SOC variation curve of SESS under 25 s continuous disturbance. (**b**) SOC variation curve of BESS under 25 s continuous disturbance.

As shown in Figure 14 and Table 3, the proposed control strategy achieves the smallest frequency fluctuation, and the conclusion is supported by the lowest RMS value of the frequency. Compared to the other three control strategies, the RMS of FD decreased by 25.57%, 33.34%, and 45.62%, respectively. Meanwhile, Figure 15a and Table 3 reveal that the SOC maintenance ability of the SESS under the proposed strategy is slightly inferior to that of the fixed-K method and adaptive VDC method, but the difference in SOC's RMS is minimal, and the maximum SOC fluctuation does not exceed 0.1, ensuring the SESS operates in a highly safe charge state. The SOC of the BESS remains consistently stable at 0.5, showing that the BESS does not need to be discharged under minor short-term disturbances. The BESS's SOC RMS further confirms that the proposed strategy adequately safeguards the SOC of the BESS. This demonstrates that the priority output of SESS significantly reduces the losses of the BESS.

This result shows that under continuous load disturbances, the proposed control strategy exhibits superiority in both frequency regulation and maintaining the SOC of the ESS.

5.4. Simulation Analysis of Combined Disturbances

To further evaluate the PFR performance of the system under long-duration disturbance conditions and simulate a more realistic power grid environment, a semi-realistic composite disturbance signal based on historical datasets, as illustrated in Figure 16, was applied to the system. The FD curve and the SOC of the ESSs are illustrated in Figures 17 and 18, respectively. The corresponding performance metrics are shown in Table 4.

Table 4. Evaluation index of PFR under 250 s continuous combined load disturbance.

Control Policies	$f_{\text{index}}/10^{-4}$ p.u.	SOC_{SCindex}	SOC_{Bindex}
Paper strategy	5.1355	0.0809	0.0458
Adaptive VDC strategy	5.9258	0.0224	0.1459
Fixed-K method	6.5755	0.0217	0.0652
No ESS	7.2071	-	-

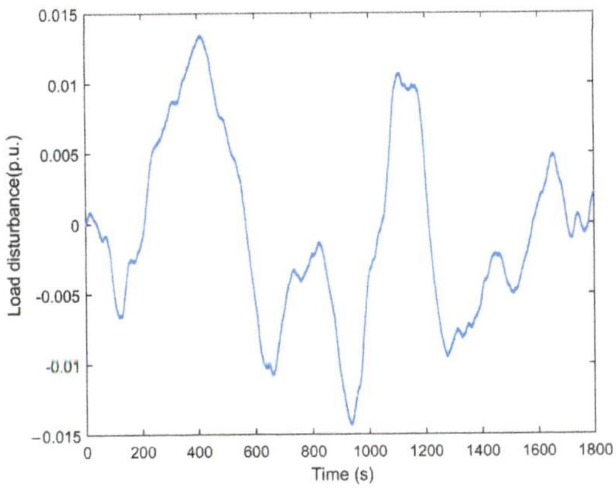

Figure 16. Variation curve of combined disturbance for 1800 s.

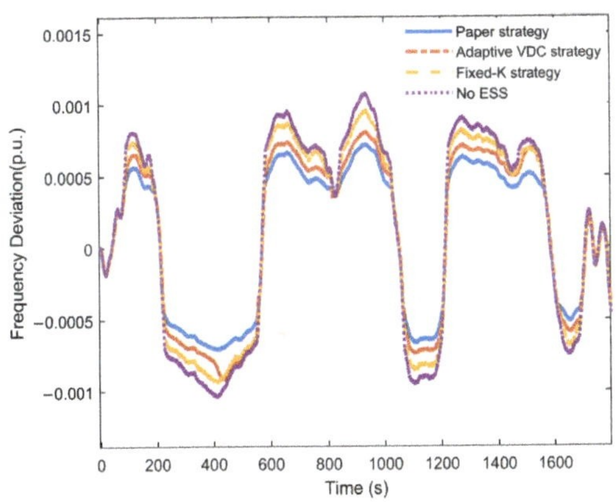

Figure 17. FD curve under 1800 s combined disturbance.

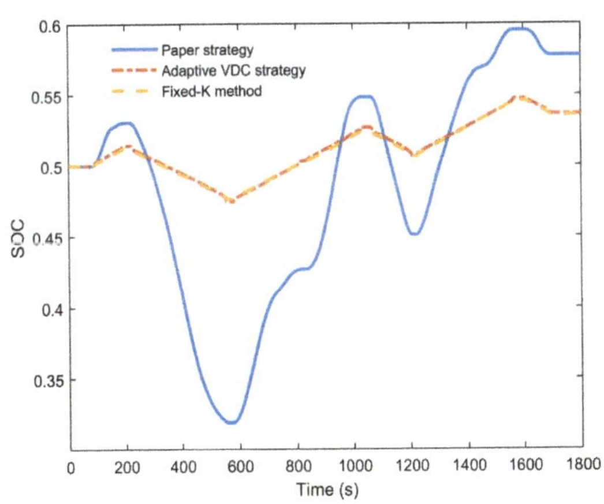

(**a**)

Figure 18. *Cont.*

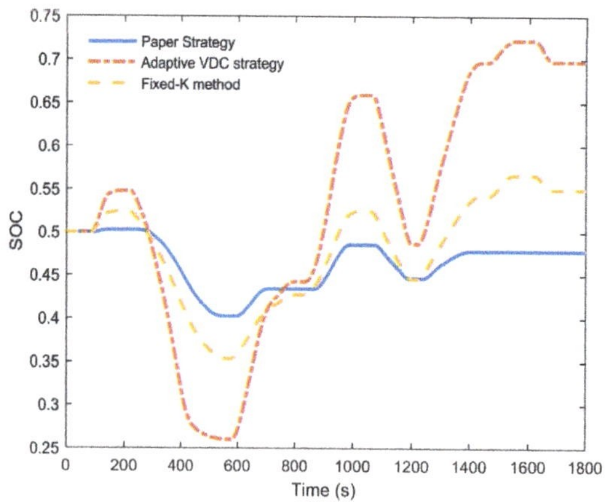

(b)

Figure 18. SOC variation curve of ESS under 1800 s combined disturbance. (**a**) SOC variation curve of SESS under 1800 s combined disturbance. (**b**) SOC variation curve of BESS under 1800 s combined disturbance.

As shown in Figure 17, the system without ESS experiences the most severe frequency fluctuations. As shown in Table 4, compared to the other three control strategies, the RMS of FD decreased by 15.39%, 21.89%, and 40.34%, respectively. This indicates that this control strategy effectively stabilizes frequency fluctuations during daily grid operations and offers strong engineering applicability.

As illustrated in Figure 18a, the proposed control strategy is less effective in maintaining the SOC of the SESS compared to the other two systems. However, the SOC fluctuations remain within the safe range, so the disadvantage is not significant. As shown in Figure 18b and Table 4, the SOC of the BESS exhibits minimal fluctuations under the combined disturbances. Compared to the other two strategies, the RMS of BESS's SOC decreased by 68.61% and 29.75%, respectively. This is attributed to the narrow dead zone of the SESS, which enables it to respond promptly and suppress most frequency fluctuations within the BESS's dead zone. Therefore, under this strategy, the SESS can independently respond to continuous small-scale disturbances most of the time, significantly reducing the frequency of BESS discharge. This enhances the lifespan of the BESS while leveraging the long cycle life of the SESS.

Combined frequency disturbances more closely resemble real-world grid conditions, and simulation results show that the proposed control strategy better maintains grid frequency while effectively preserving the SOC of the SESS and extending the life of the BESS.

5.5. Simulation Analysis of Commercial Load Disturbance

To further evaluate the system's PFR performance under commercial loads with random characteristics, a random disturbance of ± 0.005 p.u. is added to the real load shown in Figure 9, simulating a more variable commercial load. The FD curve and the SOC of the ESSs are illustrated in Figures 19 and 20, respectively. The corresponding performance metrics are shown in Table 5.

Due to the intense random fluctuations in commercial loads, the complete FD graph can not intuitively show the advantages and disadvantages of the four control strategies in terms of PFR. However, by observing the localized zoom of the FD in Figure 19 along with the RMS value of FD in Table 5, it is clear that the proposed control strategy achieves the best PFR performance under commercial loads. Compared to the other three control strategies, the RMS of FD decreased by 7.69%, 20%, and 33.33%, respectively. As shown in Figure 20a,b and Table 5, similar to the composite disturbance case, under commercial loads, the proposed control strategy's SESS lacks an advantage in maintaining SOC but can effectively maintain the SOC of the BESS. Compared to the other two strategies, the RMS of BESS's SOC decreased by 21.36% and 23.65%, respectively.

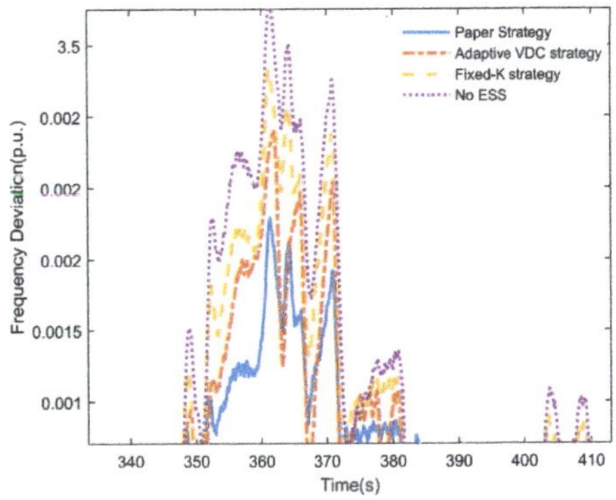

Figure 19. Local magnified curve of FD variation under commercial load.

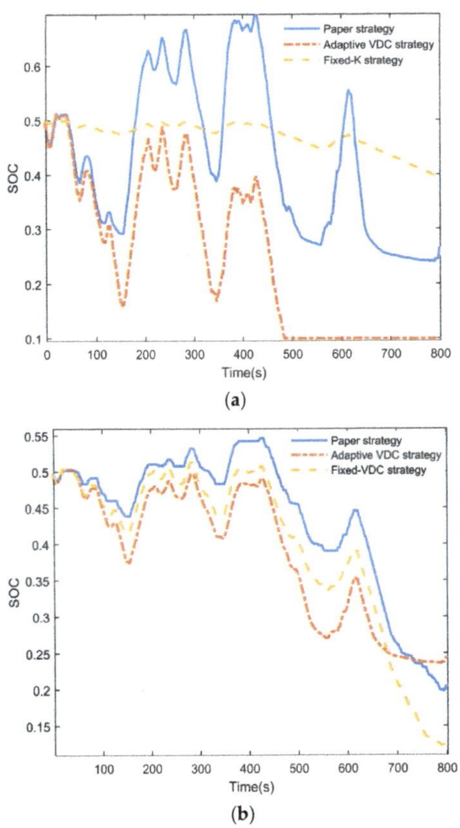

Figure 20. SOC variation curve of ESS under 800 s commercial disturbance. (**a**) SOC variation curve of SESS under 800 s commercial disturbance. (**b**) SOC variation curve of BESS under 800 s commercial disturbance.

Table 5. Evaluation index of PFR under 800 s continuous commercial disturbance.

Control Policies	f_{index}/p.u.	SOC_{SCindex}	SOC_{Bindex}
Paper strategy	0.0012	0.1606	0.1156
Adaptive VDC strategy	0.0013	0.2808	0.1470
Fixed-K method	0.0015	0.0039	0.1514
No ESS	0.0018	-	-

The performance of the FD and the SOC of the ESS demonstrates the applicability of the proposed control strategy under the random characteristics of commercial load.

5.6. Simulation Analysis of Different Capacity Configurations

To further demonstrate the scalability and adaptability of the proposed control strategy, the 1800 s combined disturbance load shown in Figure 16 was applied to models with three different capacity configurations: the original capacity ("Origin Capacity"), half of the original capacity ("Half Capacity"), and twice the original capacity ("Double Capacity"). The FD curve and the SOC of the ESSs are illustrated in Figures 21 and 22, respectively.

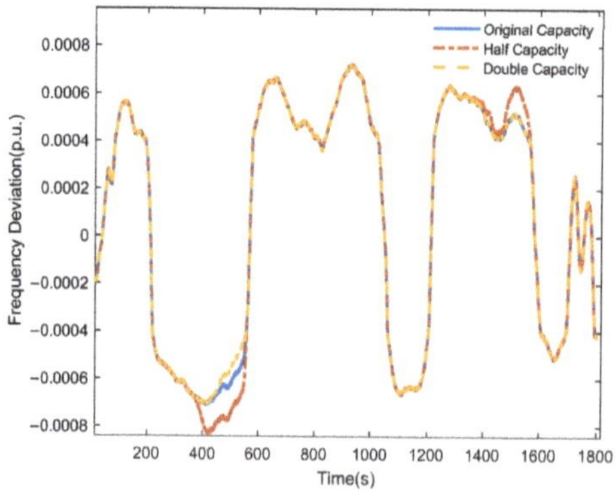

Figure 21. FD curve of different capacity configurations under 1800 s combined disturbance.

As shown in Figure 21, the proposed control strategy can effectively mitigate FD under different capacity configurations. As shown in Figure 22a,b, both the SESS and BESS are able to maintain the SOC within a reasonable range under different capacity configurations. The capacity configuration adopted by the proposed control strategy results in relatively low HESS energy consumption while achieving good frequency regulation performance and avoiding high initial investment costs, demonstrating its rationality.

5.7. Economic Analysis

The composite disturbance illustrated in Figure 15 is introduced into the systems with HESS and BESS-only primary frequency regulation, respectively. The rain flow counting method is used to calculate the number of cycles at various depths of charge and discharge, which are then converted to equivalent full cycles at a depth-of-discharge (DoD) of 1. The actual lifespan of the ESS is then estimated based on its cycle life.

In this simulation case, the BESS-alone PFR leads to 0.7254 equivalent cycles within 1800 s, yielding a BESS lifespan of about 143.61 days. After integrating the SESS, the BESS output is reduced, and the equivalent number of cycles over 1800 s drops to 0.5033, extending the BESS lifespan to 206.96 days. This significantly reduces the operating and maintenance costs of the BESS.

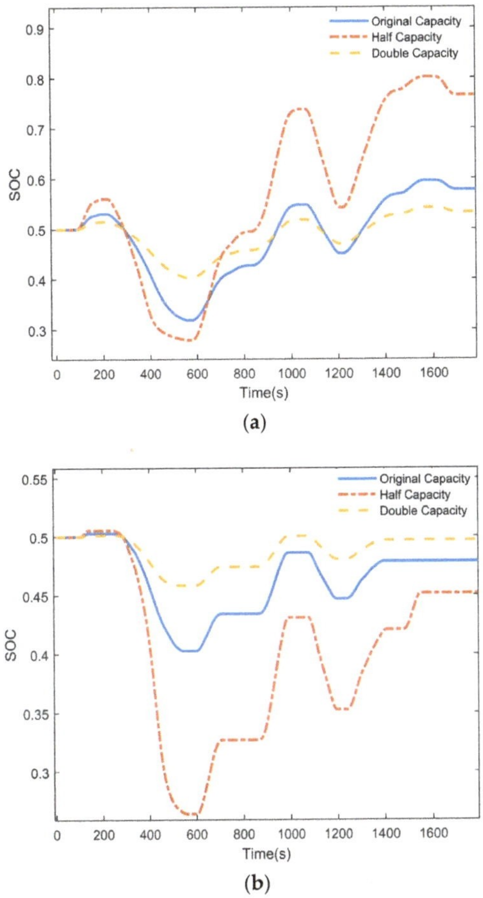

Figure 22. SOC variation curve of different capacity configurations under 1800 s combined disturbance. (**a**) SOC variation curve of SESS of different capacity configurations under 1800 s combined disturbance. (**b**) SOC variation curve of BESS of different capacity configurations under 1800 s combined disturbance.

The effectiveness of SESS in mitigating BESS capacity degradation can be qualitatively inferred from Equation (30),

$$Q_{loss} = B \cdot \exp\left[\frac{-31700 + 370.3 \times C_{rate}}{RT}\right](N_{cycle} \times DoD \times 2 \times C_{rate})^{0.55}, \quad (30)$$

where Q_{loss} represents the percentage of capacity loss, N_{cycle} is the equivalent full cycle count, DoD is the depth of discharge, C_{rate} is the rated capacity, and B, R, and T are the pre-exponential factor, gas constant, and absolute temperature, respectively.

Compared with BESS-only frequency regulation, the HESS reduces both the equivalent cycle count and DoD, thereby slowing capacity degradation and extending service life.

To verify the economic efficiency and engineering practicality of the proposed control strategy, the net present benefit P_{NET} was calculated for both the proposed strategy and the strategy where only the BESS participates in PFR. The economic parameters for the SESS and BESS are adopted from reference [59]. The calculation results are shown in Table 6.

Table 6. Economic indicators for various ESSs participating in PFR.

	BESS in HESS	BESS in HESS	BESS Only
C_{LCC}/USD	7.8426×10^7	1.728×10^7	9.857×10^7
N_{RES}/USD	2.0394×10^8	5.1281×10^8	2.9188×10^8
P_{NET}/USD	1.2551×10^8	4.9553×10^8	1.9331×10^8

As clearly illustrated in Table 6, the economic benefit of HESS participating in PFR is 3.212 times that of using only BESS.

The point at which the P_{NET} becomes zero corresponds to when the benefits outweigh the costs, which defines the economic lifetime [60]. The calculated economic lifetime is 1.4776 years for the HESS and 4.7137 years for the BESS-only system. The proposed strategy shortens the economic lifetime by approximately 3.23 years, significantly enhancing the economic viability of ESS-based frequency regulation.

5.8. Discussion on Simulation Verification

The above analysis, based on simulation experiments under various operating conditions and capacities, as well as economic evaluation, demonstrates the effectiveness and economic benefits of the proposed coordinated control strategy for PFR.

It is important to eliminate potential errors that may affect the reliability of the simulation conclusions. The four simulated strategies in this paper employ a simplified equivalent ESS model, where the ESS behavior is represented by the product of a first-order inertia element and a control gain, without considering other device parameters [61]. This simplification introduces deviations between the simulation outcomes and actual system performance. Reference [62] has demonstrated that neglecting internal battery characteristics and using a simplified equivalent model can enhance simulation efficiency while maintaining adequate accuracy. Since all four strategies adopt the same storage model configuration, the impact of model precision on the simulation results is ruled out.

Practical implementation of this control strategy should consider real-world factors such as technical limitations, sensor requirements, and controller response times.

Firstly, the design of BESS output in this paper assumes a constant rated power throughout its operation. Reference [63] points out that lithium-ion batteries experience degradation during repeated cycling, primarily in the form of capacity loss and increased internal resistance, which results in reduced power output and weakened frequency regulation performance over time. Therefore, in practical applications, frequency regulation commands should accommodate the gradual performance degradation of batteries.

Secondly, accurate measurement of system frequency and its rate of change is fundamental for both VIC and VDC. Accurate frequency tracking requires high-precision frequency sensors. In addition, applying anti-aliasing or moving average filtering prior to sampling facilitates the delivery of accurate frequency regulation signals to the HESS [64].

Moreover, controller response time may be constrained by computational load and communication delays in real systems. Therefore, optimizing the control algorithm and improving controller performance is essential to reduce response time and enhance dynamic performance [65].

6. Conclusions

This paper proposes an HESS adaptive PFR control strategy that incorporates the frequency regulation dead zone of ESS. The conclusions are outlined as follows:

(1) The proposed control combines the advantages of VDC and VIC with the distinct characteristics of power-type and energy-type storage systems. The high-power SESS is designed to respond to inertial control signals, while the high-capacity BESS re-

sponds to VDC signals. Compared to the conventional strategy, the proposed control strategy reduces the frequency drop rate by 17.43% under step disturbance. Under compound disturbances, the RMS of frequency deviation decreases by 13.34% and the RMS of BESS's SOC decreases by 68.61%. The economic benefit of this strategy is 3.212 times that of the single-energy-storage scheme. This approach maximizes the advantages of different ESSs, exhibits superior performance in PFR and economic benefits, strengthens frequency stability of the, power grid, supports greater integration of renewable energy sources, and contributes to reducing fossil fuel dependence and carbon emissions.

(2) The strategy effectively protects the lifespan of ESSs, reduces the start/stop frequency of the TPU, and reduces the costs associated with BESS replacement, disposal of retired BESS, and TPU maintenance. By appropriately setting dead zones, the SESS is prioritized for output following the BESS, with the TPU acting as a last resort. This approach leverages the long cycle life of SESS, avoiding excessive BESS charge–discharge cycles and start/stop frequency of TPU. Additionally, to prevent overcharging and over-discharging, which could irreversibly damage the storage system, the output coefficient of ESS is adaptively controlled according to the SOC. Compared with the fixed-K method, the proposed strategy better maintains the SOC of ESSs.

(3) The proposed control strategy better adapts to the impact of different loads. The inertia coefficient and the dead zone of the ESS are flexibly adjusted according to the load type, enhancing adaptability to various operating conditions.

(4) The proposed control strategy fully utilizes SOC recovery within the dead zone for the BESS. During the SOC recovery phase of the BESS, it introduces a recovery demand coefficient determined by the SOC and a recovery constraint coefficient determined by the FD. These coefficients are used to establish an SOC recovery coefficient within the frequency dead zone, ensuring effective SOC recovery while preventing secondary frequency drops. Simulations demonstrate no occurrence of secondary frequency drops, and the SOC remains in good condition.

Nonetheless, this study still has some limitations. For instance, the power and capacity of the energy storage system are configured based on empirical values, lacking a more reasonable optimization scheme to achieve the best economic performance.

At the same time, in low-inertia grids or grids with a high proportion of uncontrollable renewable energy, the frequency and depth of frequency regulation tasks for the ESS will significantly increase. This accelerates the degradation of the ESS's capacity, further restricting its long-term frequency regulation performance. If the capacity configuration of the ESS cannot meet such high-intensity frequency regulation demands, the HESS may fail to provide stable and continuous frequency regulation services under frequent fluctuations in renewable energy, which could impact the overall frequency stability of the grid. Therefore, under low inertia or high renewable energy penetration conditions, the limitation of this strategy lies in the need for higher capacity configuration.

In addition, the BESS in this paper is treated as an integrated unit during the design phase. In practice, however, numerous battery cells must be grouped together for frequency regulation. Inconsistent output among the cells during operation may lead to premature aging of certain cells, shortening the overall lifespan of the BESS, causing capacity degradation, and even posing safety risks.

Additionally, this study establishes a simplified active power–frequency control model for HESS participating in PFR, based on the concepts of area equivalence and the flat-voltage assumption. This model omits inverter voltage loops and reactive power dynamics to highlight the active power–frequency coupling and reduce simulation complexity. As a future extension, we plan to embed Q–V droop control and virtual synchronous generator

(VSG) techniques into a detailed inverter-level model. Simulations under transient load switching and three-phase fault scenarios will be conducted to quantitatively assess the impact of the proposed strategy on bus voltage sag, recovery time, and reactive power injection, thereby systematically validating the role of HESS in enhancing voltage stability and reactive power support.

Author Contributions: H.J. conceptualized the research and wrote the initial draft. K.H. supervised the study and participated in data organization. Y.L. acquired the funding. Y.L. and W.B. participated in reviewing and editing the manuscript. All authors have read and agreed to the published version of the manuscript.

Funding: This research was funded by the Postdoctoral Fellowship Program of CPSF, grant number GZC20240909.

Data Availability Statement: The original contributions presented in this study are included in the article. Further inquiries can be directed to the corresponding authors.

Conflicts of Interest: The authors declare no conflicts of interest. Although Y.L. secured funding and participated in reviewing and editing the manuscript, the funders had no role in the design of the study; in the collection, analyses, or interpretation of data; in the writing of the manuscript; or in the decision to publish the results.

Abbreviations

The following abbreviations are used in this manuscript:

HESS	hybrid energy storage system
SOC	state of charge
TPUs	thermal power units
ESSs	energy storage systems
PFR	primary frequency regulation
FD	frequency deviation
VDC	virtual droop control
VIC	virtual inertia control
BESSs	battery energy storage systems
VNIC	virtual negative inertia control
SESS	supercapacitor energy storage system
RMS	root mean square
DoD	depth-of-discharge

References

1. Yan, G.G.; Li, D.Y.; Liu, J.H.; Mao, G. A Cost Accounting Method of the Li-ion Battery Energy Storage System for Frequency Regulation Considering the Effect of Life Degradation. *Prot. Control Mod. Power Syst.* **2018**, *3*, 1–9. [CrossRef]
2. Ma, Y.X.; Hu, Z.C.; Song, Y.H. Hour-Ahead Optimization Strategy for Shared Energy Storage of Renewable Energy Power Stations to Provide Frequency Regulation Service. *IEEE Trans. Sustain. Energy* **2022**, *13*, 2331–2342. [CrossRef]
3. Dang, J.; Seuss, J.; Suneja, L.; Harley, R.G. SoC Feedback Control for Wind and ESS Hybrid Power System Frequency Regulation. *IEEE J. Emerg. Sel. Top. Power Electron.* **2014**, *2*, 79–86. [CrossRef]
4. Li, P.Q.; Tan, Z.X.; Zhou, Y.J.; Li, C.B.; Li, R.; Qi, X.Z. Secondary Frequency Regulation Strategy With Fuzzy Logic Method and Self-Adaptive Modification of State of Charge. *IEEE Access* **2018**, *6*, 43575–43585. [CrossRef]
5. Delille, G.; François, B.; Malarange, G. Dynamic Frequency Control Support by Energy Storage to Reduce the Impact of Wind and Solar Generation on Isolated Power System's Inertia. *IEEE Trans. Sustain. Energy* **2012**, *3*, 931–939. [CrossRef]
6. Chen, W.; Sun, N.; Ma, Z.C.; Liu, W.F.; Dong, H.Y. A Two-Layer Optimization Strategy for Battery Energy Storage Systems to Achieve Primary Frequency Regulation of Power Grid. *Energies* **2023**, *16*, 2811. [CrossRef]
7. Yan, L.; Jiang, P.P.; Li, W.; Xu, Y.S.; Yang, Z.T.; Zhang, B.H.; Mao, C.X. A Mathematical Model of Versatile Energy Storage System and Its Modeling by Power System Analysis Software Package. *Power Syst. Technol.* **2012**, *36*, 51–57.
8. Mu, Y.F.; Wu, J.Z.; Ekanayake, J.; Jenkins, N.; Jia, H.J. Primary Frequency Response From Electric Vehicles in the Great Britain Power System. *IEEE Trans. Smart Grid* **2013**, *4*, 1142–1150. [CrossRef]

9. Feng, Y.H. Research on Energy Storage Battery Participating in Primary Frequency Control Strategy of Power System. Master's Thesis, Hunan University, Changsha, China, 2019.
10. Knap, V.; Chaudhary, S.K.; Stroe, D.I.; Swierczynski, M.; Craciun, B.I.; Teodorescu, R. Sizing of an Energy Storage System for Grid Inertial Response and Primary Frequency Reserve. *IEEE Trans. Power Syst.* **2016**, *31*, 3447–3456. [CrossRef]
11. Yang, P.H.; Dong, X.L.; Li, Y.; Kuang, L.; Zhang, J.H.; He, B.; Wang, Y. Research on Primary Frequency Regulation Control Strategy of Wind-thermal Power Coordination. *IEEE Access* **2019**, *7*, 144766–144776. [CrossRef]
12. Gao, S.; Li, J.; Song, H.; Ding, L.; Zhang, C. An integrated thermal power-energy storage system for improving primary frequency regulation performance of thermal power units. *Power Syst. Prot. Control* **2023**, *51*, 116–125.
13. Shui, T. Research on Integrated Control Strategy Considering Hybrid Energy Storage Participation in Grid Frequency Regulation. Master's Thesis, Shanxi University, Taiyuan, China, 2023.
14. Zhang, J.C. Research on Optimal Control Strategy of Electrochemical Energy Storage System Participating in Power Grid Primary Frequency Regulation. Master's Thesis, Changsha University of Science and Technology, Changsha, China, 2022.
15. He, L.; Guo, L.Q.; Tan, Z.X.; Chen, C.Y.; Li, X.R.; Huang, J.Y.; Li, X.Y. Dynamic Performance Improving Strategy for Primary Frequency Regulation With Energy Storages in High Penetration of RESs Power System. *IEEE Access* **2024**, *12*, 139560–139572. [CrossRef]
16. Yan, L.Q.; Shui, T.; Xue, T.L.; Wang, M.; Ma, N.; Li, K.Y. Comprehensive Control Strategy Considering Hybrid Energy Storage for Primary Frequency Modulation. *Energies* **2022**, *15*, 4079. [CrossRef]
17. Liu, Y.Q.; Chen, M.R.; Xie, B.; Ban, M.F. Integrated control strategy of BESS in primary frequency modulation considering SOC recovery. *IET Renew. Power Gener.* **2024**, *18*, 875–886. [CrossRef]
18. Huang, J.Y.; Li, X.R.; Cao, Y.J.; Liu, W.J. Capacity Allocation of Energy Storage System Considering Its Action Moment and Output Depth in Rapid Frequency Regulation. *Trans. China Electrotech. Soc.* **2015**, *30*, 454–464.
19. Fang, C.X.; Tang, Y.C.; Ye, R.; Lin, Z.S.; Zhu, Z.S.; Wen, B.Y.; Ye, C.T. Adaptive Control Strategy of Energy Storage System Participating in Primary Frequency Regulation. *Processes* **2020**, *8*, 687. [CrossRef]
20. Li, X.R.; Cui, X.W.; Huang, J.Y.; Li, S.J.; Meng, Y.P. The Self-Adaption Control Strategy of Energy Storage Batteries Participating in the Primary Frequency Regulation. Trans. *China Electrotech. Soc.* **2019**, *34*, 3897–3908.
21. Lopes, J.A.P.; Almeida, P.M.R.; Soares, F.J. Using Vehicle-to-Grid to Maximize the Integration of Intermittent Renewable Energy Resources in Islanded Electric Grids. In Proceedings of the 2009 International Conference on Clean Electrical Power (ICCEP 2009), Capri, Italy, 9–11 June 2009.
22. Wu, Q.F.; Song, X.L.; Zhang, J.R.; Yu, H.F.; Huang, J.D.; Dai, H.Y. Study on Self-Adaptation Comprehensive Strategy of Battery Energy Storage in Primary Frequency Regulation of Power Grid. *Power Syst. Technol.* **2020**, *44*, 3829–3836.
23. Li, X.R.; Deng, T.; Huang, J.Y.; Li, S.J.; Yang, J.; Deng, M. Battery Energy Storage System Self-Adaptation Control Strategy in Fast Frequency Regulation. *High Volt. Eng.* **2017**, *43*, 2362–2369.
24. Ma, J.; Pan, J.; Wu, Y.; Wang, H.; Zhang, P. Improved flexible virtual inertial control considering SOC of energy storage and characteristics of frequency recovery. *Electr. Power Autom. Equip.* **2020**, *40*, 100–107.
25. Wei, W.; Bin, H.; Ai, F.Z.; Chen, Z.F.; Xu, R.J.; Ren, Y.F. Research on battery energy storage in cooperation with thermal power units to participate in system frequency regulation control strategy. *China Meas. Test* **2022**, *48*, 1–7.
26. Tian, S.J. Research on Control Strategy of Battery Energy Storage System Under Primary Frequency Regulation of Power Grid and Unbalanced Grid Voltage. Master's Thesis, North China Electric Power University, Baoding, China, 2022.
27. Zhao, X.L.; Zhang, D.H.; Gui, Y.; Li, P.; Tan, S.Y. An Integrated Control Strategy for Energy Storage Participating in Primary Frequency Regulation Considering State of Charge Constraint. *Mod. Electr. Power* **2022**, *39*, 95–103.
28. Bi, L. Optimal Control Strategy of Energy Storage Assisting Wind for Power Grid in the Primary Frequency Regulation. Master's Thesis, North China Electric Power University, Baoding, China, 2020.
29. Zhang, Z.Y.; Ding, T.; Zhou, Q.; Sun, Y.G.; Qu, M.; Zeng, Z.Y.; Ju, Y.T.; Li, L.; Wang, K.; Chi, F.D. A review of technologies and applications on versatile energy storage systems. *Renew. Sust. Energ. Rev.* **2021**, *148*, 31. [CrossRef]
30. Guerrero, J.M.; Vasquez, J.C.; Matas, J.; de Vicuña, L.G.; Castilla, M. Hierarchical Control of Droop-Controlled AC and DC Microgrids-A General Approach Toward Standardization. *IEEE Trans. Ind. Electron.* **2011**, *58*, 158–172. [CrossRef]
31. Yoon, M.; Lee, J.; Song, S.; Yoo, Y.; Jang, G.; Jung, S.; Hwang, S. Utilization of Energy Storage System for Frequency Regulation in Large-Scale Transmission System. *Energies* **2019**, *12*, 3898. [CrossRef]
32. Liu, Z.L. Research on Frequency Regulation Characteristics of Hybrid Energy Storage Assisted Heating Units. Master's Thesis, North China Electric Power University, Baoding, China, 2023.
33. Man, Y. Research on Primary Frequency Modulation Control Method of New Energy Power Station Assisted by "Flywheel +Lithium" Hybrid Energy Storage System. Master's Thesis, Shandong University, Jinan, China, 2023.
34. Fang, H.Y. Research on Control Strategies for Hybrid Energy Storage in System Frequency Regulation. Master's Thesis, Qingdao University, Qingdao, China, 2022.

35. Liu, H.; Liu, J.; Zhao, L.; Li, Y. Adaptive Control Strategy for Primary Frequency Modulation of Hybrid Energy Storage Considering State of Charge Recovery. In Proceedings of the 2022 International Conference on Energy Materials and Electrical Engineering (ICEMEE 2022), Guilin, China, 22–24 July 2022.
36. Elkasem, A.H.A.; Kamel, S.; Khamies, M.; Nasrat, L. Frequency regulation in a hybrid renewable power grid: An effective strategy utilizing load frequency control and redox flow batteries. *Sci. Rep.* **2024**, *14*, 33. [CrossRef]
37. Zhao, X.L.; Li, P.; Fu, B. Frequency Regulation Method Assisted by Energy Storage Based on Frequency Response Characteristics. *Electr. Power Eng. Technol.* **2024**, *43*, 41–49.
38. *GB/T 15945-2008*; Power Quality—Frequency Deviation for Power System. Standards Press of China: Beijing, China, 2008.
39. Sun, H.D.; Xu, T.; Guo, Q.; Li, Y.L.; Lin, W.F.; Yi, J.; Li, W.F. Analysis on Blackout in Great Britain Power Grid on August 9th, 2019 and Its Enlightenment to Power Grid in China. *Proc. CSEE* **2019**, *39*, 6183–6192.
40. Meng, Y.; Li, X.R.; Liu, X.L.; Cui, X.W.; Xu, P.A.; Li, S.J. A Control Strategy for Battery Energy Storage Systems Participating in Primary Frequency Control Considering the Disturbance Type. *IEEE Access* **2021**, *9*, 102004–102018. [CrossRef]
41. Li, L.; Liu, X.P. Study on Frequency Modulation Control Strategy of Energy Storage Auxiliary Thermal Power Unit Considering Frequency Difference Zone. *J. North China Electr. Power Univ.* **2024**, *51*, 126–133+142.
42. Wei, N.; Yan, X.; Han, X.; Ma, X.; Cui, J.; Sun, W. Research on Control Strategy of Hybrid Energy Storage System Participating in Primary Frequency Regulation of Power Grid. *J. Phys. Conf. Ser.* **2023**, *2592*, 012061. [CrossRef]
43. Li, J.H.; Guo, Z.; Yang, H.; Li, L.; Sun, B.X.; Fan, X.K. Primary Frequency Regulation Control Strategy of Energy Storage Based on Dynamic Droop Coefficient and SOC Reference. *Power Syst. Prot. Control* **2021**, *49*, 1–10.
44. Xu, Y.H.; Li, H.Q.; Qin, J.X.; Wang, K.C.; Han, J.H. Frequency Control of Hybrid Energy Storage Assisted Thermal Power Units Based on SOC State. *Autom. Appl.* **2023**, *64*, 78–83.
45. Li, P.Q.; Fu, Y.H.; Liu, X.R.; Tan, Z.Z.; Yang, B.; Han, J.Y. Control Strategy for Energy Storage Battery Participating in Primary Frequency Regulation of Power Grid Considering Ultra-short-term Load Forecasting. *Autom. Electr. Power Syst.* **2019**, *43*, 87–93.
46. Ma, Z.H.; Li, X.R.; Tang, Z.X.; Huang, J.Y.; Li, H. Integrated Control of Primary Frequency Regulation Considering Dead Band of Energy Storage. *Trans. China Electrotech. Soc.* **2019**, *34*, 2102–2115.
47. Wuraola, A.; Patel, N.; Nguang, S.K. Efficient Activation Functions for Embedded Inference Engines. *Neurocomputing* **2021**, *442*, 73–88. [CrossRef]
48. Fang, Y.; Hu, J.Y.; Liu, W.H.; Shao, Q.Q.; Qi, J.; Peng, Y.H. Smooth and Time-Optimal S-Curve Trajectory Planning for Automated Robots and Machines. *Mech. Mach. Theory* **2019**, *137*, 127–153. [CrossRef]
49. Liu, T.Y.; Cui, J.K.; Li, Y.H.; Gao, S.Y.; Zhu, M.C.; Chen, L.H. Time-Optimal Asymmetric S-Curve Trajectory Planning of Redundant Manipulators under Kinematic Constraints. *Sensors* **2023**, *23*, 3074. [CrossRef]
50. Liu, Y.P.; Tian, S.J.; Liang, H.P.; Xie, Y.; Huo, Q.D.; Tang, X.J. Control Strategy of a Battery Energy Storage System Considering SOC in Primary Frequency Regulation of Power Grid. *Power Syst. Prot. Control* **2022**, *50*, 107–118.
51. Deng, X.; Sun, W.; Xiao, H.W. Integrated Control Method of Battery Energy Storage Participating in Primary Frequency Regulation. *High Volt. Eng.* **2018**, *44*, 1157–1165.
52. Wang, Y.F.; Yang, M.C.; Xue, H.; Zhang, Y.H.; Mi, Y. Adaptive Integrated Control Strategy for Battery Energy Storage System Considering SOC in Primary Frequency Regulation. *Electr. Power Autom. Equip.* **2021**, *41*, 192–198, 219.
53. Ju, P.; Li, H.Y.; Gan, C.; Liu, Y.; Yu, Y.P.; Liu, Y.L. Analytical Assessment for Transient Stability Under Stochastic Continuous Disturbances. *IEEE Trans. Power Syst.* **2018**, *33*, 2004–2014. [CrossRef]
54. Huang, J.Y.; Li, X.R.; Chang, M.; Li, S.J.; Liu, W.J. Capacity Allocation of Battery Energy Storage System in Primary Frequency Regulation Considering Its Technical-Economic Model. *High Volt. Eng.* **2017**, *32*, 112–121.
55. Li, J.H.; Tao, H.; Mu, G.; Yan, G.; Li, C.; He, D.C. Primary frequency regulation strategy with energy storage system based on weight factors and state of charge recovery. *Autom. Electr. Power Syst.* **2020**, *44*, 63–72.
56. Liu, J. Hybrid Energy Storage for Frequency Modulation Cooperative Control of Power Grid. Master's Thesis, North China Electric Power University, Baoding, China, 2022.
57. Zhou, D.; Zou, Z.W.; Dan, Y.Q.; Wang, C.X.; Teng, C.Y.; Zhu, Y.L. An Integrated Strategy for Hybrid Energy Storage Systems to Stabilize the Frequency of the Power Grid Through Primary Frequency Regulation. *Energies* **2025**, *18*, 246. [CrossRef]
58. Fang, H.Y.; Ma, P. Adaptive Droop Control Strategy Considering Frequency Modulation Dead Band of Hybrid Energy Storage. *J. Electr. Eng.* **2021**, *16*, 213–222.
59. Zhang, Q.; Li, X.R.; Yang, M.; Cao, Y.J.; Li, P.Q. Capacity Allocation Method of Hybrid Energy Storage System for Smoothing Wind Power Fluctuations with Maximum Net Benefit. Trans. *China Electrotechn. Soc.* **2016**, *31*, 40–48.
60. Zizlavsky, O. Net Present Value Approach: Method for Economic Assessment of Innovation Projects. In Proceedings of the 19th International Scientific Conference on Economics and Management (ICEM), Riga, Latvia, 23–25 April 2014.
61. Ye, L.; Wang, K.F.; Lai, Y.N.; Chen, H.; Zhao, Y.N.; Xu, X.; Lu, P.; Jin, Y.F. Analysis of Frequency Characteristics and Control Strategies of Battery Energy Storage Frequency Regulation in Power Systems Under Low Inertia. *Power Syst. Technol.* **2023**, *47*, 446–464.

62. Huang, J.Y.; Li, X.R.; Cao, Y.J.; Zhang, Q.; Liu, W.J. Simulation Model of Battery Energy Storage System for Grid Frequency Regulation. *Autom. Electr. Power Syst.* **2015**, *39*, 20–24, 74.
63. Vermeer, W.; Mouli, G.R.C.; Bauer, P. A Comprehensive Review on the Characteristics and Modeling of Lithium-Ion Battery Aging. *IEEE Trans. Transp. Electrif.* **2022**, *8*, 2205–2232. [CrossRef]
64. Deng, X.Y.; Mo, R.; Wang, P.L.; Chen, J.R.; Nan, D.L.; Liu, M.Y. Review of RoCoF Estimation Techniques for Low-Inertia Power Systems. *Energies* **2023**, *16*, 3708. [CrossRef]
65. Jabari, M.; Ekinci, S.; Izci, D.; Bajaj, M.; Zaitsev, I. Efficient DC Motor Speed Control Using a Novel Multi-Stage FOPD(1 + PI) Controller Optimized by the Pelican Optimization Algorithm. *Sci. Rep.* **2024**, *14*, 22442. [CrossRef] [PubMed]

Disclaimer/Publisher's Note: The statements, opinions and data contained in all publications are solely those of the individual author(s) and contributor(s) and not of MDPI and/or the editor(s). MDPI and/or the editor(s) disclaim responsibility for any injury to people or property resulting from any ideas, methods, instructions or products referred to in the content.

energies

MDPI

Article

ANN-Based Maximum Power Tracking for a Grid-Synchronized Wind Turbine-Driven Doubly Fed Induction Generator Fed by Matrix Converter

Mohamed A. Alarabi and Sedat Sünter *

Electrical and Electronics Engineering Department, Atilim University, Ankara 06830, Türkiye; malarabi1991@gmail.com or alarabi.maarelbi@student.atilim.edu.tr
* Correspondence: sedat.sunter@atilim.edu.tr

Abstract: The integration of renewable energy sources, such as wind power, into the electrical grid is essential for the development of sustainable energy systems. Doubly fed induction generators (DFIGs) have been significantly utilized in wind energy conversion systems (WECSs) because of their efficient power generation and variable speed operation. However, optimizing wind power extraction at variable wind speeds remains a major challenge. To address this, an artificial neural network (ANN) is adopted to predict the optimal shaft speed, ensuring maximum power point tracking (MPPT) for a wind energy-driven DFIG connected to a matrix converter (MC). The DFIG is controlled via field-oriented control (FOC), which allows independent power output regulation and separately controls the stator active and reactive power components. Through its compact design, bidirectional power flow, and enhanced harmonic performance, the MC, which is controlled by the simplified Venturini modulation technique, improves the efficiency and dependability of the system. Simulation outcomes confirm that the ANN-based MPPT enhances the power extraction efficiency and improves the system performance. This study shows how wind energy systems can be optimized for smart grids by integrating advanced control techniques like FOC and simplified Venturini modulation with intelligent algorithms like ANN.

Keywords: wind energy; WECS; MPPT; DFIG; FOC; MC; Venturini algorithm

1. Introduction

In the last two decades, wind energy has definitively been recognized as an applicable source of renewable energy systems, fundamentally due to its characteristic of being an inexhaustible source that could be converted to electrical energy across numerous systems called WECSs. Recently, researchers have increasingly concentrated on the optimization of WECSs to take advantage of the full potential of this abundant and non-polluting energy source [1]. Furthermore, nowadays, energy experts are looking for alternative efficacious and environmentally friendly energy resources in order to decrease the dependence on fossil fuels such as hydrocarbon- and petroleum-based fuels, the primary contributor to the problem of global warming, which are becoming scarce [2]. The increasing integration of the production of wind energy in modern power systems has brought about new difficulties and challenges. One of the significant requirements for WECSs is their capability for generating stable and desirable power in spite of fluctuating and varying wind speed situations; moreover, they react swiftly to sudden wind speed variations [3]. Numerous control schemes have been suggested for both grid-side converters and machine-side converters, interconnected through a DC bus capacitor. Additionally, WECSs use a range of electrical generator types, such as permanent magnet synchronous generators (PMSG) and DFIG, to effectively convert wind energy into electrical energy [4]. Fixed-speed and variable-speed WECSs are the two main categories into which WECSs have been divided according to their operational speed. In fixed-speed WECSs, power converters are not required because of the constancy in the speed of the generator's rotor and the direct connection to the grid [5]. In the variable-speed WECSs, in order to make sure that the turbine runs at its maximum power, the generator's shaft speed needs to be continuously modified and adjusted. As a result, the maximum power extraction (MPE) occurs, which would be the cause of the rising annual production of the energy by 5–10% [6]. Taking into consideration that the wind turbine operating regions have been divided into four operation regions presented in Figure 1, the first region encompasses the area from start-up until the cut-in point. The second region, commonly referred to as the MPPT region, spans between the cut-in and the rated speed, where the turbine's output power ought to be maximized. The third region, known as the speed regulation region, aims to keep the shaft speed at rated levels in order to ensure that the turbine does not generate power exceeding the generator's rated capacity. The fourth region occurs if the speed of the wind exceeds the cut-out point, ensuring protection against potential damage from excessively high wind speeds [2,4]. It is remarkable to note that the maximum power from WECSs can be taken out by operating the turbine in the second region [7].

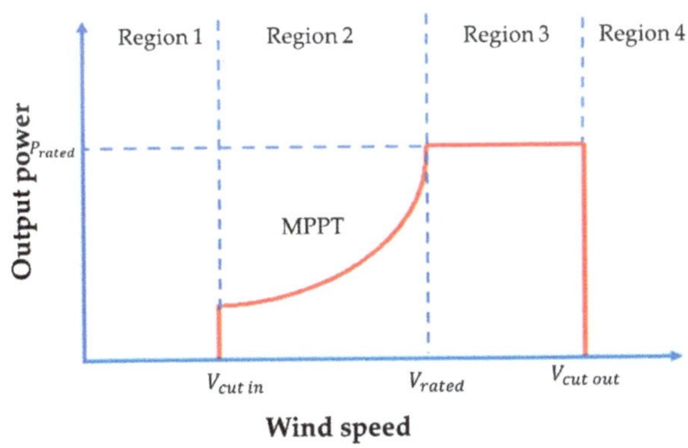

Figure 1. The wind turbine's operational regions.

To optimize the efficiency and performance, several algorithms have been applied in WECSs. Among them is the MPPT algorithm, which is widely utilized in sustainable energy systems, such as photovoltaic systems and wind energy, to maximize power extraction under varying environmental conditions [8]. In WECSs, the MPPT algorithm's primary objective is to gather the greatest possible amount of power from the wind. The two primary methods, into which MPPT algorithms have been separated in the published works, are direct power control (DPC) and indirect power control (IPC) [9]. DPC directly measures the generated output power and modifies system parameters to guarantee that the turbine performs its MPE. In contrast, the IPC technique pre-calculates the power using wind-speed data [3,8,9].

WECSs, equipped with variable-speed wind turbine-driven DFIGs, have been becoming widespread day by day due to their capability of operating throughout a broad spectrum of fluctuating wind speeds and can be operated at both sub-synchronous and super-synchronous speeds. Moreover, they can be controlled with a partial-scale converter since the converter addresses only the slip power, so that the converter components can be smaller and less expensive compared to those in full-power conversion systems. Typically, a gearbox is used to operate the DFIG, creating a link between the wind turbine and the DFIG [10–12].

WECSs are characterized by an environmentally friendly nature and low-cost installation, yet they often suffer from low efficiency because of continuously altering wind speeds [13]. Therefore, the MPPT plays a critical role in enhancing WECSs' performance. Various significant studies have been conducted to improve efficient MPPT controllers; many of them have been applied to variable speed DFIGs, which can continuously become acclimatized to changing wind speeds. This acclimation has been leading to an enhancement in the overall turbine efficiency and limited power fluctuations [14]. MPPT's fundamental objective is to maximize the power coefficient's value (Cp) to its highest value, even with fluctuations in wind speed. The existing literature covers two broad methods of MPPT strategies, which are conventional methods and the soft computing-based approach [2]. Hill climbing search (HCS), which has been also known as perturbation and observation (P&O), is one of the conventional techniques that has been documented in [15]. One more method is the optimal torque control (OTC) technique, and it is utilized to maximize the power for a WECS-based PMSG [16]. The advantages of this technique are higher efficiency and straightforwardness. However, it suffers from dependency on the climatic conditions [9]. Latterly, many researchers have been interested in soft computing approaches, divided into two categories: nature-inspired techniques and artificial intelligence (AI) techniques [2]. Ant colony optimization (ACO) [17], particle swarm optimization (PSO) [18], and genetic algorithms (GAs) [19] are a few of the nature-inspired methods adopted into the MPPT controller. Compared to conventional methods, the nature-inspired methods have demonstrated fast tracking under changeable wind speeds. However, the requirements of the multiparameter-like selection of chromosomes, crossover rate, and population size have made them complex tasks [20]. The AI-based MPPT controller in WECSs involves a fuzzy logic (FL) controller [21,22] and an ANN-based controller [23,24]. These methods do not require accurate mathematical modeling even though they operate with variable inputs and have the ability of self-convergence in addition to self-learning capabilities. Moreover, they are adaptable to the systems' non-linear behavior. The FL technique has been used for maximizing the extracted power in WECSs [21,22]. However, its tracking performance and efficiency depend on the predefined rule base, which decreases its practical applicability. In addition, it requires a large scale of hypothetical knowledge and might not ensure the best possible response [2]. In [23], the ANN technique was utilized for pitch angle controllers to enhance the power maximization from the available wind in a grid-connected wind turbine system. The outcomes demonstrated that the controller-based ANN has better performance in contrast against traditional strategies.

In [25], in order to control DFIG-based-WECS, MPPT control based on the adaptive neuro-fuzzy inference (ANFI) method was employed. The simulation results of the ANFI-based approach were compared with those obtained using a classical PI controller. The frequently used method in WECSs is the sliding mode control (SMC) method. It offers robustness with variations in system parameters and can effectively handle limited external disturbances. Thus, it has been widely utilized to control the squirrel cage induction generator [26]. It has been reported in [27] that the fractional-order PI controller method has been utilized for controlling the pitch angle, and the simulation results demonstrated an enhancement in the capacity of DFIG at strong wind conditions. Other control methods have been also reported in the literature, such as the proportional integral [28], the linear-quadratic-Gaussian control method [29], and quantitative feedback theory [30], which have been used for the MPPT of WECSs. However, such methods seek inclusive computational and graphical analysis and because they are not robust, they often result in oscillations at output power [31].

It is essential for variable-speed WECSs to convert power from a source with varying speed into electrical energy with a consistent frequency. In a variable-speed WECS that uses DFIG, the main grid is directly connected to the stator side windings, so that fixed-frequency electric power can be generated through the stator side, whereas the windings of the rotor are supplied with inconstant voltage and frequency. To integrate the rotor-side with the main grid, a controllable power electronic converter is required [10]. Additionally, effective speed control is required to achieve MPT [11].

Grid-connected variable-speed WECSs driven by DFIG typically employ a back-to-back converter that enables bidirectional power flow [32]. However, back-to-back converters involve two-stage power conversion, which might require a complex control algorithm to control the entire system effectively. Moreover, the large DC link capacitors which are used in the converter system may increase volume, weight, and cost [33].

The DFIG can be controlled by a direct AC-AC MC in place of a traditional back-to-back converter. The MC has a number of positive effects, including the ability to guarantee sinusoidal input, output currents, and bidirectional power flow in a single-stage conversion process without the necessity for a DC link capacitor. It has also gained popularity because of its unity input power factor. Unlike traditional back-to-back converters, the MC has a simpler control strategy and does not seek large and expensive elements for energy storing. Furthermore, it can extend the speed range of the DFIG to be operated over synchronous speed, so that the DFIG can deliver power not only from the stator but also from the rotor to the main grid through MC. All the above-mentioned advantages have made the MC a perfect choice for the wind power generation systems [34,35]. Most control strategies for wind turbine-driven DFIG systems are typically based on the FOC approach [10,36].

In DFIG-based WECSs, the FOC schematic is specifically designed to regulate the rotor currents in such a way that the power can flow bidirectionally from and to the grid, depending on the rotor speed and operation conditions. This approach allows the shaft speed to follow the reference value, which is determined by the curves of power-speed characteristic of the wind turbine [10]. A block diagram of a grid-connected DFIG-based WECS with MC is illustrated in Figure 2.

The ANN-based MPPT strategies have been discussed for several renewable energy systems in the literature. However, no prior study has combined an ANN-based MPPT approach with a MC-fed DFIG for a WECS. The proposed control strategy, which has been presented in [10], uses a 2D lookup table for achieving the MPPT in a DFIG-based WECS with a MC. While effective, the 2D lookup table method seeks offline data and lacks adaptation to unexpected operating conditions. To address these limitations, this study proposes replacing the 2D lookup table with an ANN-based MPPT approach, offering improved adaptability, dynamic response, and reduced reliance on offline-generated data. This paper presents the design, simulation, and evaluation of the proposed ANN-based MPPT strategy.

The main contributions of this study are summarized as follows: proposing an ANN-based MPPT approach to replace the conventional 2D lookup table utilized in reference [10]; integrating the ANN-based MPPT with the FOC strategy in a DFIG-based WECS fed by an MC, which ensures bidirectional power flow with unity power factor; developing a global simulation model in MATLAB/Simulink (version: R2024a)to evaluate the proposed system under different wind conditions; and providing a comparative analysis of the ANN and the 2D lookup table approach from [10].

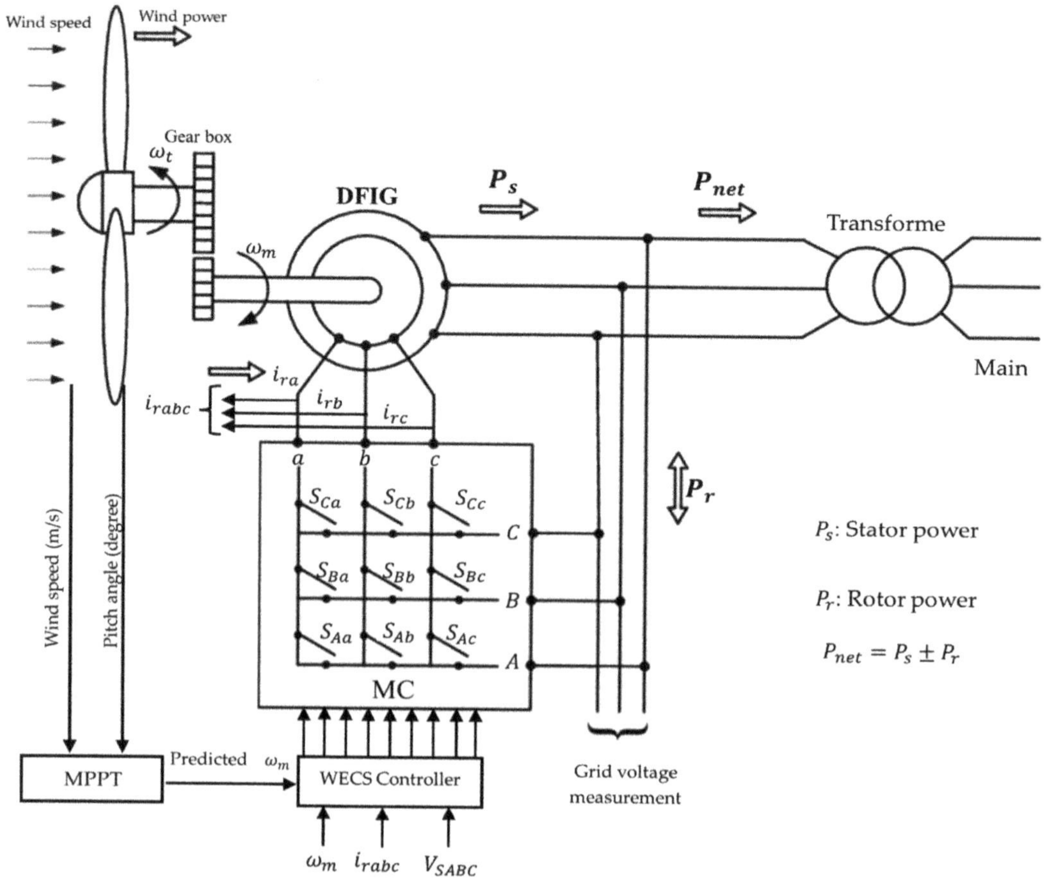

Figure 2. The diagram of a variable-speed WECS.

2. Wind Turbine Model

The produced power and mechanical torque, both of which depend on the airflow across the blades and are significantly influenced by varying wind speeds, are represented by the wind turbine's aerodynamic model [3]. The mechanical power on the turbine's rotor is influenced by the power coefficient $C_p(\lambda, \theta)$, which varies with wind speed (v_w) in $\frac{m}{s}$ and blade pitch angle (θ) in degrees, can be expressed as follows [10,37]:

$$P_w = \frac{1}{2} C_p(\lambda,\ \theta)\rho\ \pi\ R^2\ v_w^3 \tag{1}$$

where R is the turbine rotor-plane radius (m); ρ is the air density $\left(\frac{Kg}{m^3}\right)$; and λ is the tip speed ratio, which is calculated by Equation (2):

$$\lambda = \frac{\omega_t R}{v_w} \tag{2}$$

Equation (3) gives the generated mechanical torque by the turbine:

$$T_t = \frac{P_w}{\omega_t} \tag{3}$$

where ω_t is the turbine shaft's speed in $\frac{rad}{s}$. The mechanical coupling between the generator and turbine is commonly achieved by a gearbox, whose ratio G is calculated as Equation (4):

$$G = \frac{\omega_m}{\omega_t} \tag{4}$$

where ω_m is the generator speed $\frac{rad}{s}$. If the power obtained from the wind turbine overrides the generator's rated power, limiting the wind turbine's input power is essential and pitch angle control achieves this. However, when the obtained power in a range below the generator-rated power, the turbine ought to extract the maximum amount of power [10,38]. It should be noted that the power coefficient $C_p(\lambda, \theta)$ is at its highest level when the pitch angle value is zero.

$$C_p(\lambda,\,\theta) = 0.22\left(\frac{116}{\lambda_i} - 0.4\,\theta - 5\right) e^{-\frac{12.5}{\lambda_i}} \tag{5}$$

$$\frac{1}{\lambda_i} = \frac{1}{\lambda + 0.08\,\theta} - \frac{0.035}{1+\theta^3} \tag{6}$$

As illustrated in Figure 3, the output power of the turbine is shown remarkably and simultaneously rising as the speed of the wind rises. It is also obvious that the curve of the output power attains its peak value at a definite shaft speed. Therefore, the turbine should be run at the speed value that maximizes the power. Keep in mind that the value of C_p, which is a non-linear function of tip speed ratio λ and pitch angle θ, differs depending on the system. Theoretically, the highest value of C_p is 0.44, as depicted in Figure 4 [10]. The mathematical details of C_p is found in [39].

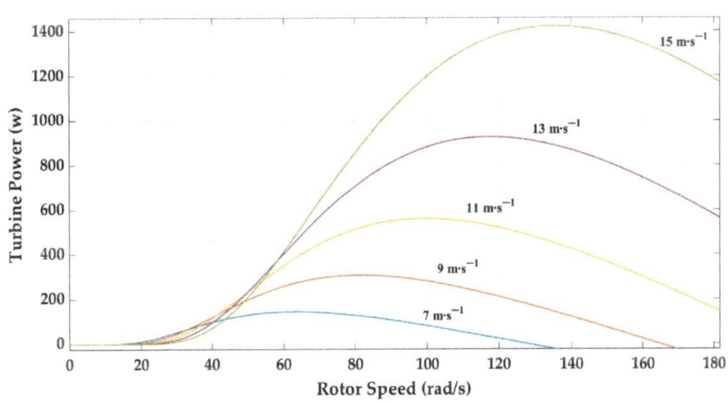

Figure 3. The turbine characteristics in terms of power and speed for varying wind speeds.

The wind turbine characteristics shown in Figures 3 and 4 were drawn based on Equations (2), (5) and (6).

In this paper, the pitch angle control not only regulates the aerodynamic power input at high wind speeds but also acts as an inherent rotor speed-limiting mechanism by reducing the aerodynamic torque. As a result, there was no need for a separate speed limiter in the control system. This interaction was embedded in the training data of the ANN, where the pitch angle values reflect both normal operation and power-limiting conditions. Accordingly, the ANN was trained to predict the optimal shaft speed based on both wind speed and pitch angle.

The wind turbine model represents a laboratory-scale turbine that has been designed to align with the characteristics of the 1 kW DFIG, which is available in the laboratory, ensuring compatibility for the simulation and potential experimental validation.

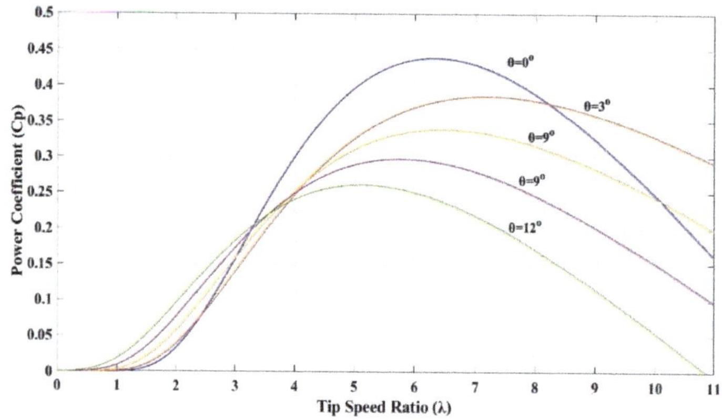

Figure 4. The turbine's characteristics at different wind speeds.

3. The MC and Simplified Form of Venturini Algorithm

The MC, which consists of bidirectional switches, functions as a direct AC-AC converter. It has the ability to convert an input voltage with a constant amplitude and frequency to a variable output voltage at different frequencies, and it eliminates the requirement of the intermediate DC link capacitor. Consequently, it provides a substitute solution to the conventional back-to-back converter, which is ordinarily and traditionally utilized to efficiently control the DFIG-based WECSs. The array of nine bi-directional switches, made up from semiconductor materials, establishes a direct connection between the three-phase source and the three-phase load, forming the direct three-phase AC-AC conversion. Figure 5 illustrates the schematic representation of MC where three groups of bi-directional switches have been coordinated; each group is connected with an output line. This coordination would provide a direct connection between the input and output phases [40–42]. Due to the direct supply of the MC from source voltage, it is imperative to keep away from short circuiting the input phase. Additionally, the output phase ought never to be open circuited if the load is inductive. For the MC to be operated safely, these two guidelines are necessary [42].

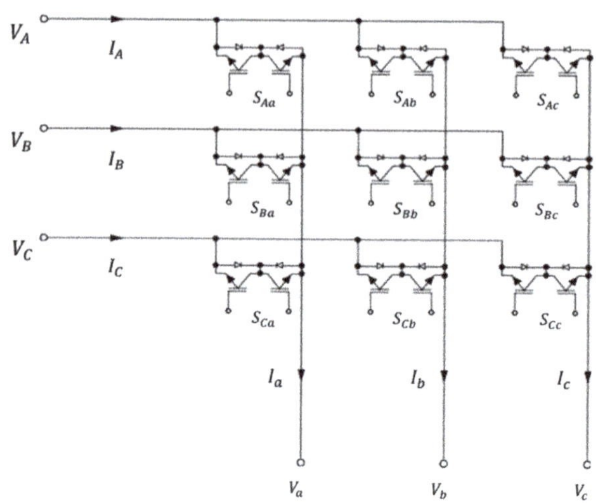

Figure 5. The power circuit for the MC.

In this study, a simplified form of the Venturini modulation algorithm (Sunter–Clare algorithm) [43], has been used to control the MC. Implementation of this algorithm is easier and more appropriate for closed-loop operations. It is supposed that the MC is supplied by a balanced three-phase constant voltage source. The input voltages v_A, v_B, v_C, and modulation terms can be used to represent the MC output voltages v_a, v_b, v_c as in Equation (7).

$$\begin{bmatrix} v_a \\ v_b \\ v_c \end{bmatrix} = \begin{bmatrix} M_{Aa} & M_{Ba} & M_{Ca} \\ M_{Ab} & M_{Bb} & M_{Cb} \\ M_{Ac} & M_{Bc} & M_{Cc} \end{bmatrix} \begin{bmatrix} v_A \\ v_B \\ v_C \end{bmatrix} \quad (7)$$

To use this modulation technique, at least two of the three input line-to-line voltages need to be measured. Then, the peak value of input voltages V_{im} and its position ω_{it} can be expressed as Equation (8):

$$V_{im}^2 = \frac{4}{9}\left[V_{AB}^2 + V_{BC}^2 + V_{AB}V_{BC}\right] \quad (8)$$

$$\omega_i t = atan\left(\frac{V_{BC}}{\sqrt{3}\left(\frac{2V_{AB}}{3} + \frac{V_{BC}}{3}\right)}\right) \quad (9)$$

where V_{AB}, V_{BC} are the line voltages. The target voltage's peak magnitude, V_{om} and its position, ω_{ot} can be calculated by Equations (10) and (11):

$$V_{om}^2 = \frac{2}{3}\left[V_a^2 + V_b^2 + V_c^2\right] \quad (10)$$

$$\omega_o t = arctan\left(\frac{V_b - V_c}{\sqrt{3}\,V_a}\right) \quad (11)$$

where the target phase output voltages are denoted by V_a, V_b, andV_c. The voltage magnitude and angle are directly derived from the output of the control loops. Equation (12) is used to obtain the voltage ratio q:

$$q = \sqrt{\frac{V_{om}^2}{V_{im}^2}} \quad (12)$$

Note that the maximum value of q cannot exceed 0.866. Equations (13)–(15) are used to obtain the modulations for the output phase, a:

$$M_{Aa} = \frac{1}{3} + K_{31} + \frac{2}{3V_{im}^2}(V_a + K_{33})\left(\frac{2V_{AB}}{3} + \frac{V_{BC}}{3}\right) \quad (13)$$

$$M_{Ba} = \frac{1}{3} + K_{32} + \frac{2}{3V_{im}^2}(V_a + K_{33})\left(\frac{V_{BC}}{3} - \frac{V_{AB}}{3}\right) \quad (14)$$

$$M_{Ca} = 1 - (M_{Aa} + M_{Ba}) \quad (15)$$

The triple harmonics K are injected into the target output voltage in order to achieve the maximum voltage ratio [44], and can be found using Equations (16)–(18):

$$K_{31} = \frac{2q}{9q_m}\sin(\omega_i t)\sin(3\omega_i t) \quad (16)$$

$$K_{32} = \frac{2q}{9q_m}\sin\left(\omega_i t - \frac{2\pi}{3}\right)\sin(3\omega_i t) \quad (17)$$

$$K_{33} = -\sqrt{V_{om}^2}\left[\frac{1}{6}\cos(3\omega_o t) - \frac{1}{4q_m}\cos(3\omega_i t)\right] \quad (18)$$

where q_m is the highest value of transfer ratio which equals to 0.866.

Based on Equations (13)–(18), for the unity power factor, the turn-on times of the switch locate between the input phase and the output phase are expressed as Equation (19).

$$T_{\beta\gamma} = T_s\left[\frac{1}{3} + \frac{2V_{o\gamma}V_{i\beta}}{3V_{im}^2} + \frac{2q}{9q_m}\sin\left(\omega_i t + \varnothing_\beta\right)\sin(3\omega_i t)\right] \tag{19}$$

where $V_{o\gamma}$ and $V_{i\beta}$ are the output and input voltages and can be calculated as follows:

$$V_{o\gamma} = qV_{im}.\cos\left(\omega_o t + \varnothing_\gamma\right) - \frac{q}{6}V_{im}\cos\left(3\omega_o t\right) + \frac{1}{4}\frac{q}{q_m}V_{im}\cos(3\omega_i t),\ V_{i\beta} = V_{im}\cos\left(\omega_i t + \varnothing_\beta\right) \tag{20}$$

where $\varnothing_\gamma$, $\varnothing_\beta = 0, \frac{2\pi}{3}, \frac{4\pi}{3}$.

In Equations (13) and (14), V_a is substituted with V_b and V_c, respectively, to yield the modulation functions for phases b and c. It should be mentioned that the target output waveforms in Equation (10) do not have to be strictly sinusoidal. Equation (19) can be used to obtain the MC's input currents.

$$\begin{bmatrix} I_A \\ I_B \\ I_C \end{bmatrix} = \begin{bmatrix} M_{Aa} & M_{Ba} & M_{Ca} \\ M_{Ab} & M_{Bb} & M_{Cb} \\ M_{Ac} & M_{Bc} & M_{Cc} \end{bmatrix} \begin{bmatrix} I_a \\ I_b \\ I_c \end{bmatrix} \tag{21}$$

Switching signals for the MC were produced using a simplified version of the Venturini modulation method. The MC provides the controlled rotor voltage and frequency required to carry out FOC and decoupled active and reactive power regulation of the DFIG, taking the role of the traditional rotor-side back-to-back converter in this system. The MC supports the active and reactive control techniques used in this work as well as the MPPT by allowing bidirectional power flow and variable-frequency operation at the rotor side.

MC Model

MC has been modeled in MATLAB/Simulink as a switching-based model employing nine bidirectional switches coordinated in a 3 × 3 matrix arrangement. The FOC strategy has generated reference voltage signals to synthesize the required voltage and frequency for the rotor side. The simulation model operates the MC at a switching frequency of 5 kHz. Figure 6 illustrates the implementation of one phase of MC in Simulink. The other two phases are not shown for clarity since they are the same except for a phase shift of $\frac{2\pi}{3}$ and $\frac{4\pi}{3}$ [44]. Ideal switches have been assumed in the simulation of the MC circuit. Figure 7 shows in detail the switching period calculation of phase (a) using a simplified form of Venturini algorithm.

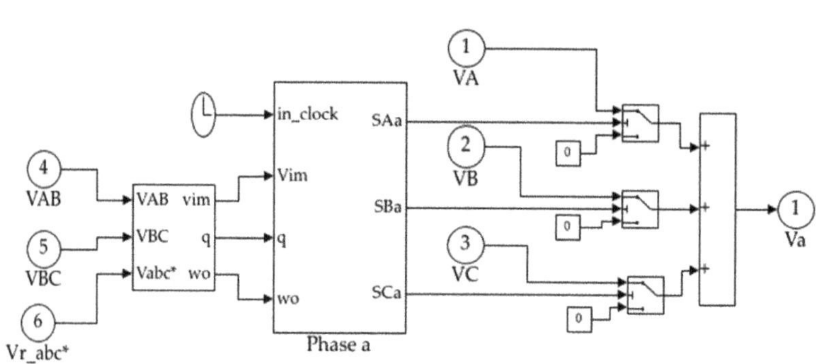

Figure 6. Detail of the Simulink block diagram of the MC for one phase.

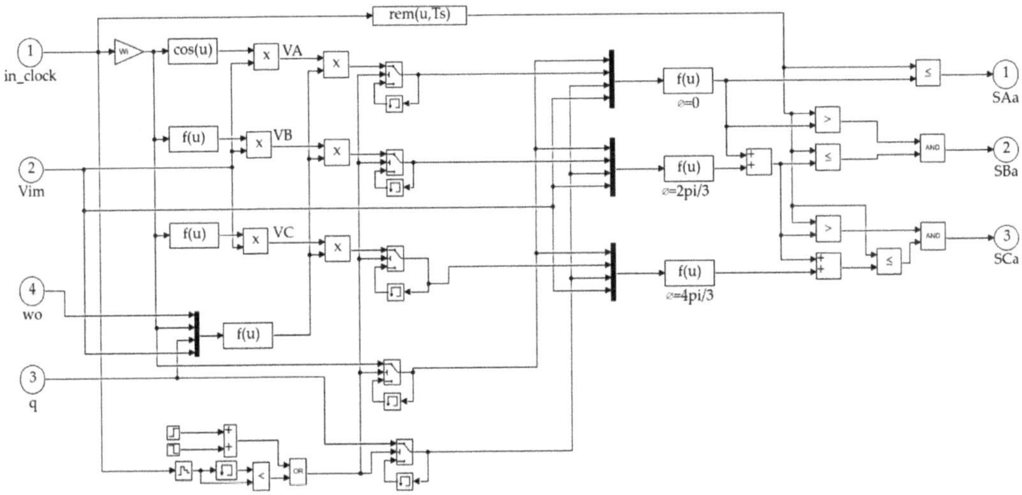

Figure 7. Switching period calculation of phase (a) using simplified form of Venturini algorithm.

In the Simulink block diagrams "*" refers to the reference values.

4. Dynamic Model of DFIG

The DFIG equivalent circuit in the $d-q$ frame is displayed in Figure 8. All the machine parameters are referred to the stator side. This approach simplifies the machine structure and allows us to easily simulate the machine in closed-loop control systems [3].

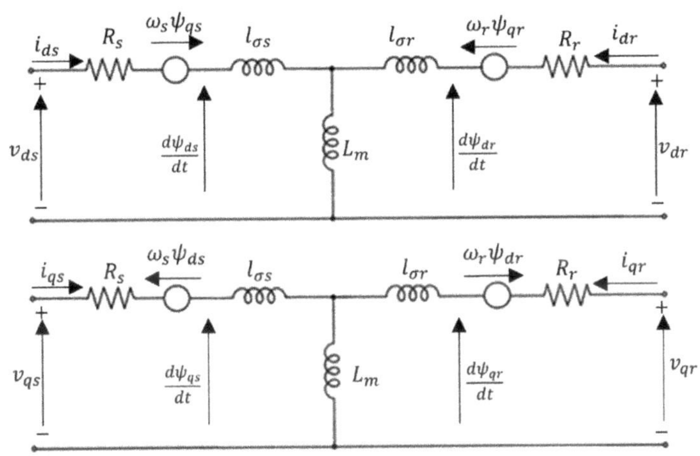

Figure 8. The equivalent circuit of DFIG in $d-q$ frame.

Equations (22)–(29) represent the DFIG model in the synchronously rotating $d-q$ reference frame [16]:

$$V_{sd} = R_s I_{sd} + \frac{d\psi_{sd}}{dt} - \omega_e \psi_{sq} \tag{22}$$

$$V_{sq} = R_s I_{sq} + \frac{d\psi_{sq}}{dt} + \omega_e \psi_{sd} \tag{23}$$

$$V_{rd} = R_r I_{rd} + \frac{d\psi_{rd}}{dt} - (\omega_e - \omega_r)\psi_{sq} \tag{24}$$

$$V_{rq} = R_r I_{rq} + \frac{d\psi_{rq}}{dt} + (\omega_e - \omega_r)\psi_{sd} \tag{25}$$

$$\psi_{sd} = L_s I_{sd} + L_m I_{rd} \tag{26}$$

$$\psi_{sq} = L_s I_{sq} + L_m I_{rq} \tag{27}$$

$$\psi_{rd} = L_r I_{rd} + L_m I_{sd} \tag{28}$$

$$\psi_{rq} = L_r I_{rq} + L_m I_{sq} \tag{29}$$

By rearranging the previous equations with the flux linkages taken into consideration as state variables, it is simple to obtain state space representation of the $d-q$ expression as Equation (30).

$$\frac{d}{dt}\begin{bmatrix}\psi_{ds}\\ \psi_{qs}\\ \psi_{dr}\\ \psi_{qr}\end{bmatrix} = \begin{bmatrix}\frac{-R_s}{\sigma L_s} & \omega_e & \frac{R_s L_m}{\sigma L_s L_r} & 0\\ -\omega_e & \frac{-R_s}{\sigma L_s} & 0 & \frac{R_s L_m}{\sigma L_s L_r}\\ \frac{R_s L_m}{\sigma L_s L_r} & 0 & \frac{-R_r}{\sigma L_r} & \omega_r\\ 0 & \frac{R_s L_m}{\sigma L_s L_r} & -\omega_r & \frac{-R_r}{\sigma L_r}\end{bmatrix}\begin{bmatrix}\psi_{ds}\\ \psi_{qs}\\ \psi_{dr}\\ \psi_{qr}\end{bmatrix} + \begin{bmatrix}v_{ds}\\ v_{qs}\\ v_{dr}\\ v_{qr}\end{bmatrix} \tag{30}$$

If currents are utilized as state–space variables rather than fluxes, the DFIM's state-space model in the synchronous reference frame will be as Equation (31).

$$\frac{d}{dt}\begin{bmatrix}i_{ds}\\ i_{qs}\\ i_{dr}\\ i_{qr}\end{bmatrix} = \left(\frac{1}{\sigma L_s L_r}\right)\begin{bmatrix}-R_s L_r & \omega_r L_m^2 + \omega_e \sigma L_s L_r & R_r L_m & \omega_r L_m L_r\\ -\omega_r L_m^2 - \omega_e \sigma L_s L_r & -R_s L_r & -\omega_r L_m L_r & R_r L_m\\ R_s L_m & -\omega_r L_m L_r & -R_r L_s & -\omega_r L_m^2 + \omega_e \sigma L_s L_r\\ \omega_r L_m L_r & R_s L_m & \omega_r L_m^2 - \omega_e \sigma L_s L_r & -R_r L_s\end{bmatrix}$$

$$\begin{bmatrix}i_{ds}\\ i_{qs}\\ i_{dr}\\ i_{qr}\end{bmatrix} + \left(\frac{1}{\sigma L_s L_r}\right)\begin{bmatrix}L_r & 0 & -L_m & 0\\ 0 & L_r & 0 & -L_m\\ -L_m & 0 & L_s & 0\\ 0 & -L_m & 0 & L_s\end{bmatrix}\begin{bmatrix}v_{ds}\\ v_{qs}\\ v_{dr}\\ v_{qr}\end{bmatrix} \tag{31}$$

The previous arrangement is used to represent the DFIG in $d-q$ frame since it is helpful to obtain steady state for given stator and rotor input voltages.

The electrical torque and mechanical dynamics of the system are being represented using Equations (32) and (33), respectively.

$$T_e = 3\frac{P}{2}L_m\left(I_{sq}I_{rd} - I_{sd}I_{rq}\right) \tag{32}$$

$$J\frac{d\omega_m}{dt} = T_e + T_{Load} - B\omega_m \tag{33}$$

where s and r represent stator and rotor quantities, respectively. L_m, L_s, and L_r are magnetizing inductance, stator, and rotor self-inductances, respectively. T_e and T_{Load} denote the electrical and load torques. B is the friction coefficient, and P indicates the pole number. ω_m denotes the mechanical speed in $\frac{rad}{s}$, and J stands for moment of inertia. Figure 9 illustrates the Simulink model of the DFIG in $d-q$ frame.

By utilizing the quadrature component of the rotor current, the DFIG dynamic model permits control of the electromagnetic torque $\left(I_{rq}\right)$. This control mechanism enables the system to follow the reference speed generated by the MPPT algorithm while keeping active and reactive power under stator-flux FOC decoupled.

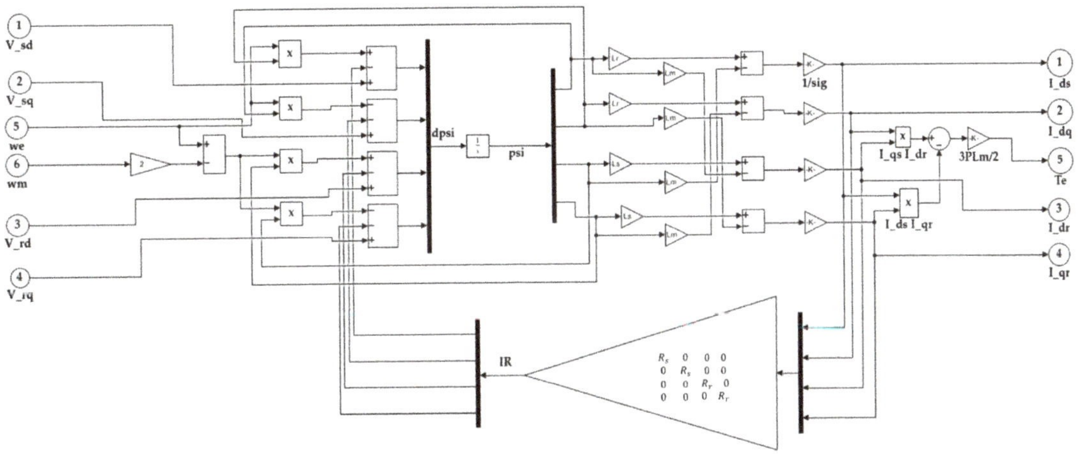

Figure 9. DFIG's Simulink model.

5. Prediction of the Shaft Speed

For the estimation of the non-linear mapping between the independent variables, which are the speed of the wind speed (V_w), the blade pitch angle (θ), and the dependent variable, which is chosen to be the turbine's shaft speed (ω_m), is the parameter whose power is to be maximized, an ANN is utilized. In neural networks, the training set for this non-linear estimation or prediction implicates the states V_w and θ as inputs and ω_m as an output. A three-layer ANN is an efficacious technique for nonlinear function prediction. Figure 10 displays this study procedure's schematic diagram.

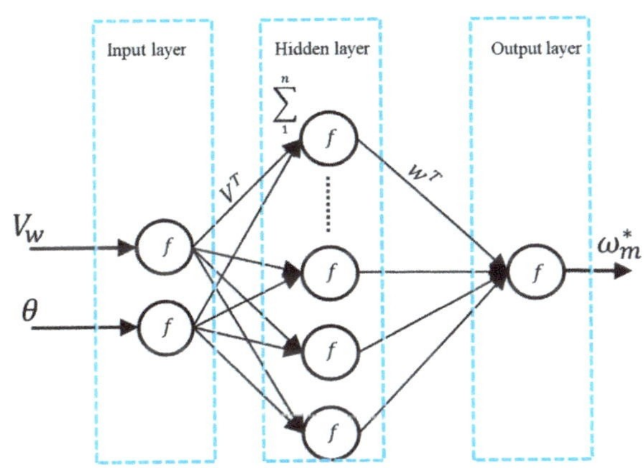

Figure 10. Neural network for predicting shaft speed.

It is essential to note that the pitch angle in this system functions not only as an aerodynamic control variable but also as a power-limiting mechanism. During the high wind, when the generator power exceeds the rated power, the pitch angle is regulated to limit the turbine's input power and indirectly reduce rotor speed. Therefore, the ANN training dataset inherently captures both normal operation and power-limited scenarios through variations in θ. By including this interaction in the training data, the ANN can accurately predict the optimal rotor speed under both unconstrained and power-limited conditions, which accordingly eliminates the need for a separate speed limiter

To calculate the net activation a_j in a neural network, data are received by the input layer, multiplied by the relevant weights V_{ji}, and then added to a bias term b_{jo}. Equation (34) expresses the calculation of the net activation of the input layer:

$$a_j = \sum_{i=1}^{l_o} V_{ji}p_i + b_{jo} \tag{34}$$

where p_i is the input to the ith node; b_{jo} is the associated bias term; and $j = 1,2,3\ldots lo$ denotes the hidden layer's number of neurons. An activation function f applied to the net activation value yields the hidden layer's output, which is determined by the following:

$$y_i = f(a_j) = f\left(\sum_{i=1}^{l_o} V_{ji}p_i + b_{jo}\right) \tag{35}$$

where f represents the activation function, selected to be hyperbolic tangent $(tanh)$.

Then, the net activation of the output layer is calculated as Equation (36):

$$a_k = \sum_{j=1}^{l_o} W_{kj}y_i + b_{ko} \tag{36}$$

where k represents the neuron number of the output layer, and W_{kj} represents the weight, which has a scalar value, between the jth node in the hidden layer and the kth node at the output layer. Based on its net activation, the output layer generates the desired shaft speed, ω_m^* as output:

$$\omega_m^* = f_1(a_k) \tag{37}$$

As expressed in Equation (38), the weights relating to the input and hidden layers V_{ji}, and the weights relating to the hidden and output layers W_{kj}, can all be used to indicate the output of the model.

$$\omega_m^* = f_1\left(\sum_{j=1}^{l_o} W_{kj}f\left(\sum_{i=1}^{n} V_{ji}p_i + b_{jo}\right) + b_{ko}\right) \tag{38}$$

This expression can be expressed in vector form as Equation (39):

$$\omega_m^* = \overline{f}\left(\overline{W}^T\overline{f}\left(\overline{V}^T\overline{p} + b_v\right) + b_w\right) \tag{39}$$

This equation can be stated more specifically as Equation (40):

$$\omega_m^* = (W tanh(V\overline{p} + b_v) + b_w) \tag{40}$$

Once the network structure has been chosen, the mean squared error (MSE), commonly referred to as the cost function, is typically defined as Equation (41):

$$J\left(V_{ji}, w_{kj}\right) = \frac{1}{2}\sum_{i=1}^{i_o}(\omega_m - \omega_m^*)^2 \tag{41}$$

where ω_m the actual optimal shaft speed and ω_m^* is the ANN-predicted shaft speed.

Results from ANN Simulation

The weights of the ANN model have been modified using the Levenberg–Marquardt training procedure. Either the maximum permitted number of iterations or the MSE standard determines when the iterative process ends. To estimate an accurate shaft speed based on the training data, a range of network parameter values, such as the number of neurons distributed across the hidden layers, have been methodically changed. The flowchart of ANN model is given in Figure 11.

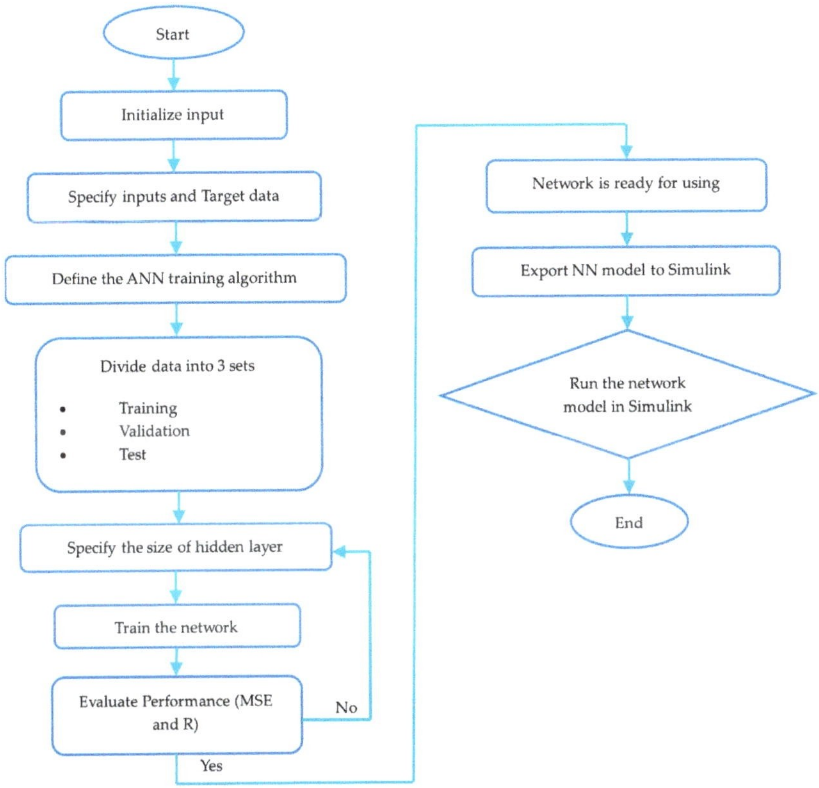

Figure 11. Flow chart of ANN.

The training data were generated by simulating a wind turbine model designed to match the 1 kW DFIG. A total of 6001 samples were generated by varying wind speed from 3 m/s to 15 m/s and pitch angle from 0° to 12°. For each combination, the optimal rotor speed was calculated by maximizing the power coefficient Cp. The dataset was randomly divided into 70% training, 15% validation, and 15% testing. ANN parameters are listed in Table 1.

Three layers make up the final ANN structure for ω_m prediction. The input variables are found in the first layer, also referred to as the input layer, contains the input variables V_w and θ. The second layer, refers to the hidden layer, contains 15 neurons. The third layer, known as the output layer, predicts the shaft speed ω_m^* that maximizes the generated power. With this selection of network parameters, the predicted and actual values match excellently.

There is a massive error at the beginning of the prediction; however, as demonstrated in Figure 12, the error decreases as the number of epochs increases. The regression plot of the ANN model has been illustrated in Figure 13. The regression value R determines if the prediction is successfully performed or needs to be trained again. It is obviously seen that R = 1, which is the optimal value of the prediction process.

Table 1. ANN parameters.

Neuron Network Architecture	Multi-Layer Perceptron Feedforward
Inputs	Pitch angle and wind speed
output	Turbine Shaft speed
Number of neurons at each layer input Hidden Output	2 15 1
Training function (Algorithm)	(Levenberg–Marquardt algorithm)

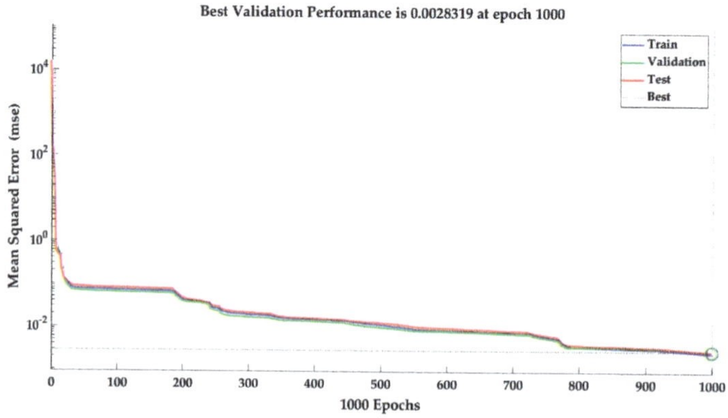

Figure 12. Performance of the ANN.

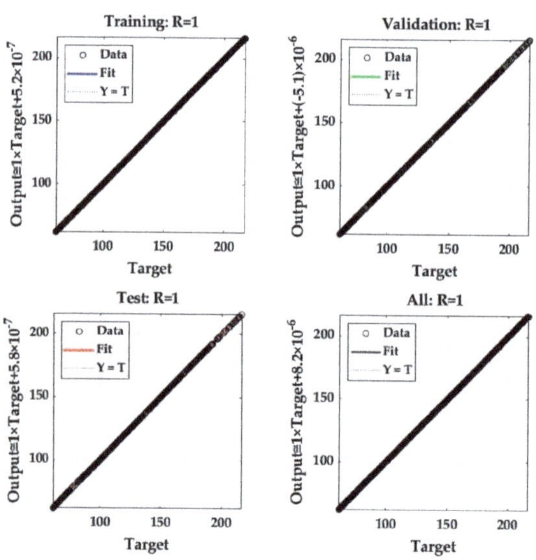

Figure 13. The regression plot of ANN.

Figure 14 displays the error histogram related to the shaft speed prediction. It displays very little inaccuracy with almost zero average value. The predicted shaft speed, which has been obtained so far, will be used as the speed reference for controlling the DFIG.

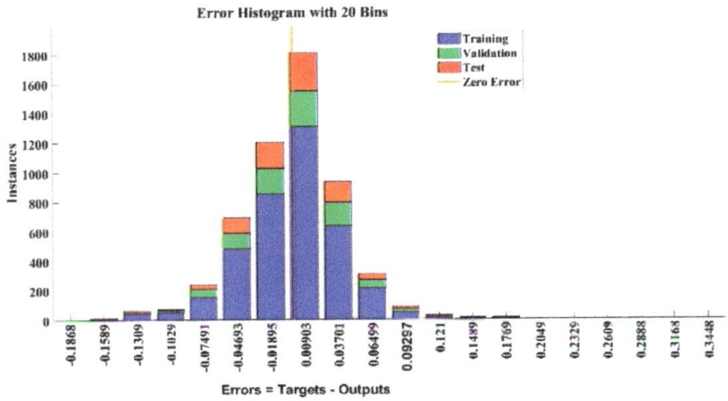

Figure 14. The errors in the histogram plot of ANN.

6. Field-Oriented Controller

The FOC method has been implemented together with the MC modulation technique for controlling the DFIG. The control has been executed in a synchronously rotating reference frame, where the flux through the windings of the stator has been aligned with the $d-axis$.

$$\psi_{sq} = 0 \text{ and } \frac{d\psi_{rq}}{dt} = 0, \psi_{sd} = L_m I_{ms} \text{ and } \frac{d\psi_{rq}}{dt} = 0$$

where I_{ms} is DFIG's stator magnetizing current.

By assuming the $q-axis$ component of the stator flux to be zero, the flux is directed entirely along the d-axis. In this configuration, the rotor current's q component controls active power control, while its d component manages reactive power control. This approach will enhance the system's dynamic performance and efficiency [10].

The substantiality of the proposed control approach can be acquired by measuring the stator voltage, rotor current, and rotor mechanical speed. Then, the stator magnetizing current can be estimated as Equation (42):

$$\frac{\tau_{ms}}{L_m} V_{sd} + I_{rd} = I_{ms} + \tau_{ms} \frac{dI_{ms}}{dt} \tag{42}$$

The stator's time constant, τ_{ms} is given by $\frac{L_s}{R_s}$. The electrical angular velocity can be estimated by Equation (43):

$$\omega_e = \frac{\frac{T_{ms}}{L_m} V_{sq} + I_{rq}}{T_{ms} I_{ms}} \tag{43}$$

$$V_{rd} = R_r I_{rd} + \sigma L_r \frac{dI_{rd}}{dt} + V_{rdc} \tag{44}$$

$$V_{rq} = R_r I_{rq} + \sigma L_r \frac{dI_{rq}}{dt} + V_{rqc} \tag{45}$$

$$T_e = -3 \frac{P}{2} \frac{L_m^2}{L_s} I_{ms} I_{rq} \tag{46}$$

$$P_s = -3\omega_e \frac{L_m^2}{L_s} I_{ms} I_{rq} \tag{47}$$

$$Q_s = 3\omega_e \frac{L_m^2}{L_s} I_{ms} (I_{ms} - I_{rd}) \tag{48}$$

where σ is the leakage coefficient; P_s is the real power of the stator; and Q_s is the reactive power of the stator. Then, compensation terms, V_{rdc} and V_{rqc}, are expressed as Equations (49) and (50):

$$V_{rdc} = (1-\sigma)L_r\frac{dI_{ms}}{dt} - (\omega_e - \omega_r)\sigma L_r I_{rq} \tag{49}$$

$$V_{rqc} = (\omega_e - \omega_r)\sigma L_r I_{rd} + (\omega_e - \omega_r)(1-\sigma)L_r I_{ms} \tag{50}$$

The leakage coefficient is expressed by Equation (51).

$$\sigma = \frac{L_s L_r - L_m^2}{L_s L_r} \tag{51}$$

These are the equations used in the control system. Block diagram representation of the wind turbine-driven DFIG employing FOC with an ANN-based MPPT algorithm are shown in Figure 15.

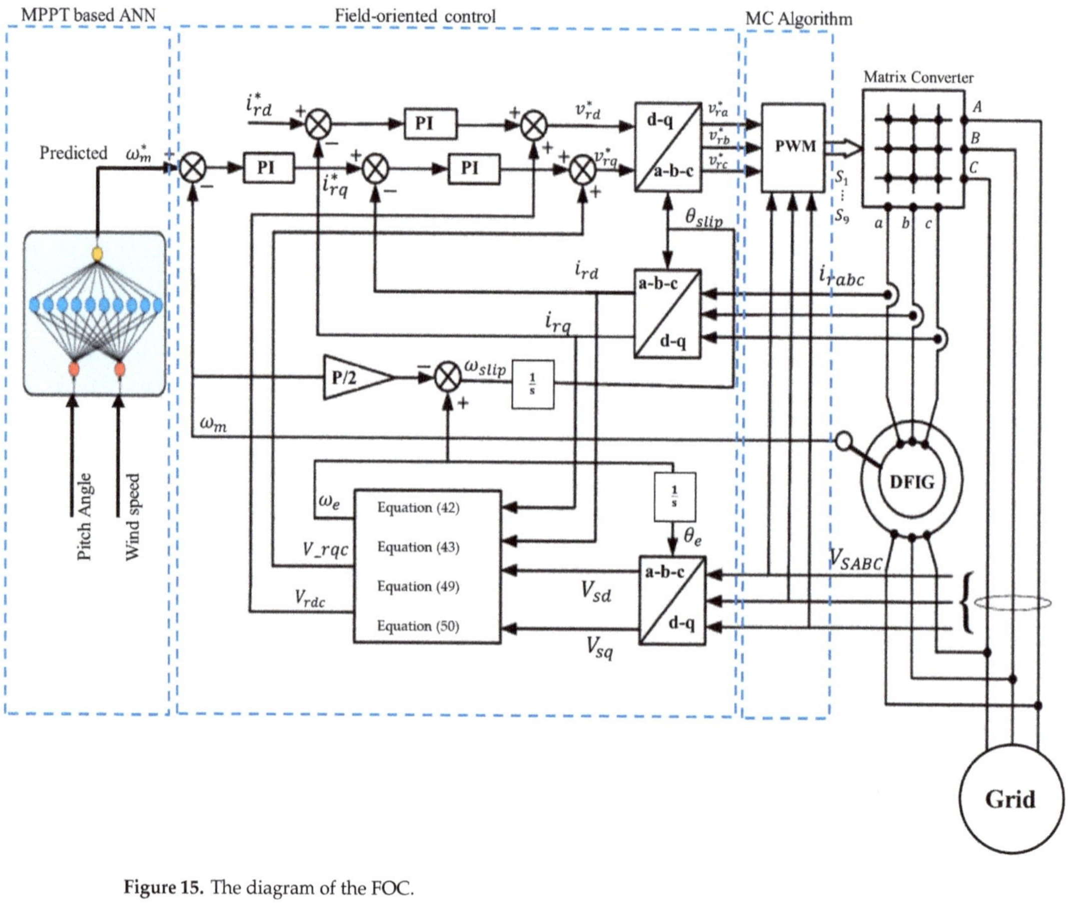

Figure 15. The diagram of the FOC.

By synthesizing the regulated rotor voltage and frequency directed by FOC method, the MC acts as the rotor-side converter, achieving decoupled active and reactive power control in the DFIG, as seen in Figure 15.

7. Simulation Results

Simulations based on the MATLAB/Simulink environment have been carried out to assess the effectiveness and performance of the suggested control system for overall system, and the entire system model in Simulink blocks is displayed in Figure 16. The objective is to validate the impact of the ANN-based MPPT and the FOC scheme in maximizing power extraction and ensuring stable operation under variable wind conditions. The parameters of the DFIG were taken from an actual machine used in the laboratory. In order to ensure the compatibility and realistic simulation performance, the wind turbine parameters were determined to match the operational characteristics of the DFIG. Turbine and DFIG parameters are displayed in Table 2.

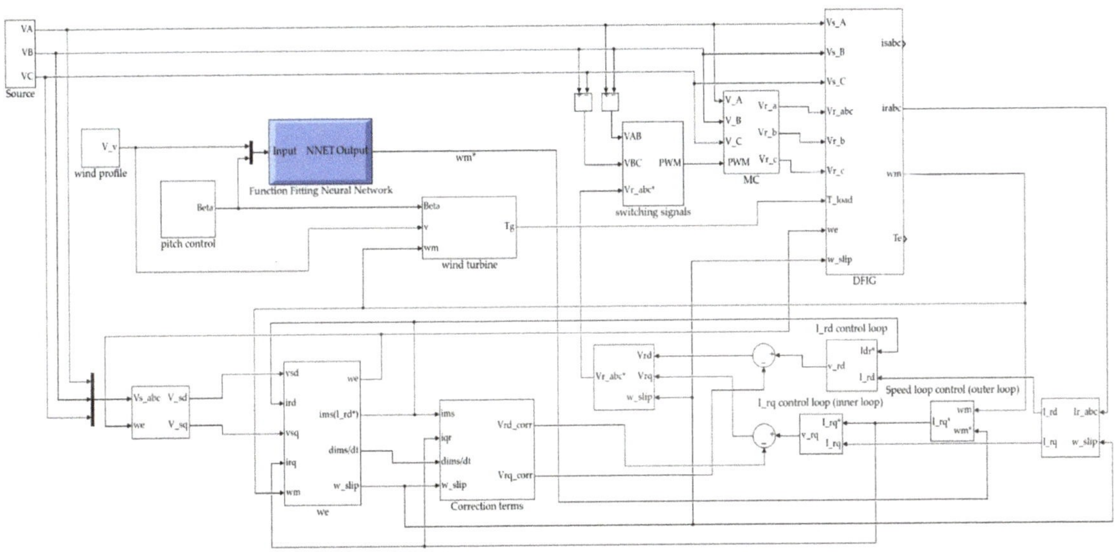

Figure 16. MATLAB/Simulink model of the entire system.

The main objective of the simulation was to assess the ANN-based MPPT performance under various wind speed situations during regular grid operation without replicating grid failures or transient disruptions.

The WECS model-based DFIG with MC has been operated at wind speeds that fluctuate over time using a pitch-angle control mechanism. Unless the power produced by the wind exceeds the generator's rated power, the value of the pitch-angle value continues to be zero. Otherwise, it would be regulated. Through the adopted control strategy, the turbine was able to achieve its optimal power by tracking the maximum value of C_p when the pitch angle is zero degrees as demonstrated in Figure 17. However, at the intervals between 22.4 and 37 s, the pitch angle was regulated due to the excess wind power generated by the turbine. This pitch angle regulation protects the turbine from severe wind powers by adjusting the C_p value.

The ANN model, which predicts or forecasts the ideal shaft speed that optimizes the power, gives the possible wind speeds and pitch angle values based on the curves of the power and speed of the wind turbine. Wind turbine shaft speed is transmitted to the generator side through the gearbox ratio. The predicted speed, which guarantees ideal tracking and MPE, is utilized as the reference speed for the DFIG's rotor. Figure 18 shows how the DFIG functions at sub-synchronous and super-synchronous speeds. The stator windings of the four-pole DFIG are connected to a 50 Hz constant voltage source. The generator operates at super-synchronous region when the rotor speed exceeds 1500 rpm. Otherwise, it operates at sub-synchronous region.

Table 2. The wind turbine and DFIG parameters.

Parameters	Symbol	Value	Unit
Turbine Parameters			
Length of the blade	R	0.7	m
Air density	ρ	1.25	k_g/m^3
Gearbox ratio	G	1.4	-
DFIG Parameters			
The nominal power	P	1000	W
The number of pole pairs	p	2	-
The stator resistance	R_s	9.83	Ω
The stator inductance	L_s	0.0292	H
The rotor resistance	R_r	8.14	Ω
The rotor inductance	L_r	0.0292	H
The mutual inductance	L_m	0.4294	H
Inertia	j	0.01	$k_g m^2$
Viscous friction coefficient	B	0.005	N·m·s

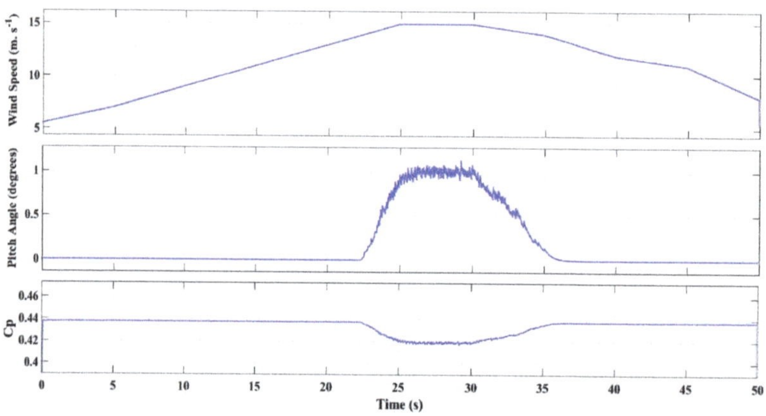

Figure 17. The power coefficient C_p, pitch angle, and wind speed profile.

DFIG power is divided into stator and rotor components. While the stator's real power P_s is always negative, indicating that it is being supplied to the main grid, the rotor's real power P_r, which depends on the rotor speed, can be either supplied to or drawn from the main grid. In Figure 19, it is seen that under sub-synchronous conditions, the power of the rotor has a positive sign, indicating that the power is being absorbed by the rotor, whereas at the super-synchronous condition, the power has a negative value, indicating the power is being delivered to the main grid. The algebraic sum of both powers will give the net power generated by the machine $P_{net} = P_s \pm P_r$, which is presented in Figure 19.

The stator's reactive power curve is illustrated in Figure 20. It has been controlled to remain zero due to the assumption that the power factor should be unity; this can be achieved according to Equation (48) by setting the reference of the current I_{rd} to be equal to the magnetizing stator current I_{ms}. However, in cases where reactive power is needed to be delivered to the main grid, the current I_{rd} will be increased to be more than the magnetizing current; this is performed by multiplying the reference current by a factor that makes it above the magnetizing current I_{ms}. The reference value of q component of rotor current I_{rq} is obtained from the output of speed controller loop.

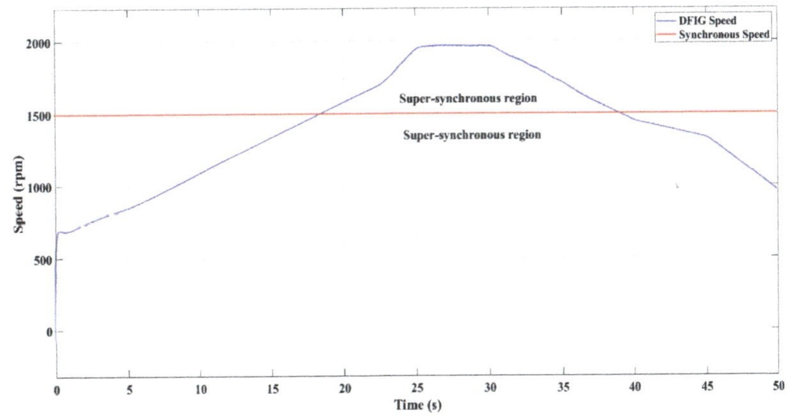

Figure 18. The speed of DFIG.

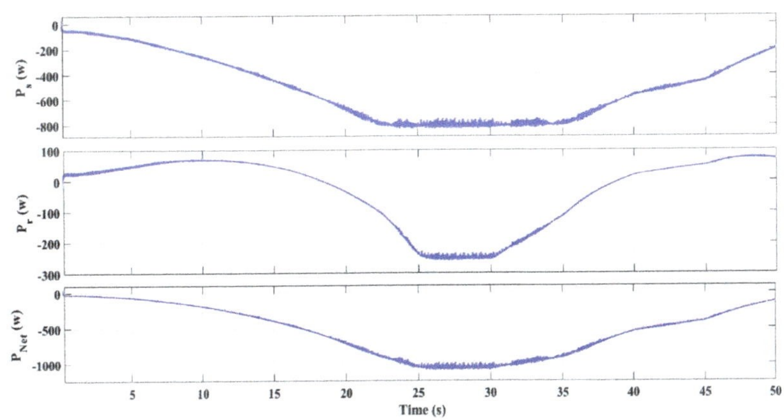

Figure 19. The DFIG stator power P_s, rotor power P_r, and net power P_{net}.

In Figure 21, as long as the stator's reactive power is kept at zero, there will be a phase angle of 180° between the voltage of the main grid and DFIG's stator current, indicating the pure real power exchange by the stator side, ensuring that it is operating with power factor of unity. Additionally, despite the variation in the rotor speed, the stator current frequency stays fixed at 50 Hz because it is connected to the main grid directly.

Figure 22 demonstrates the waveforms of the MC input current and the grid phase voltage during sub-synchronous operation region of the DFIG. During this region, the waveforms will be in phase, which indicates that the rotor of the machine is absorbing power from the main grid. However, in Figure 23, it has been illustrated that while the DFIG operates in the super-synchronous region, there will be a phase shift of 180° between the voltage and current, indicating the power is being transmitted from the rotor to the grid over MC.

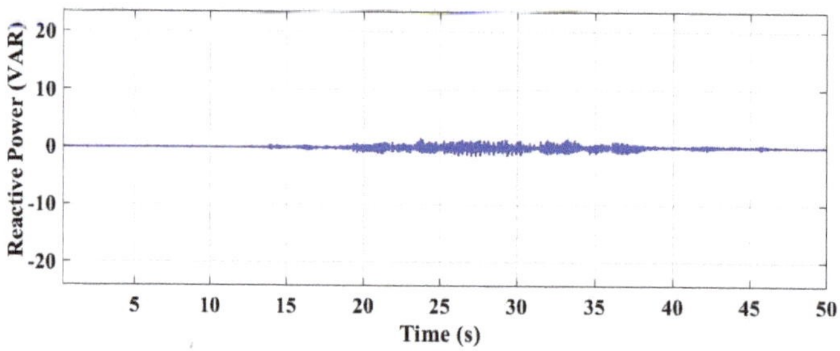

Figure 20. The stator reactive power.

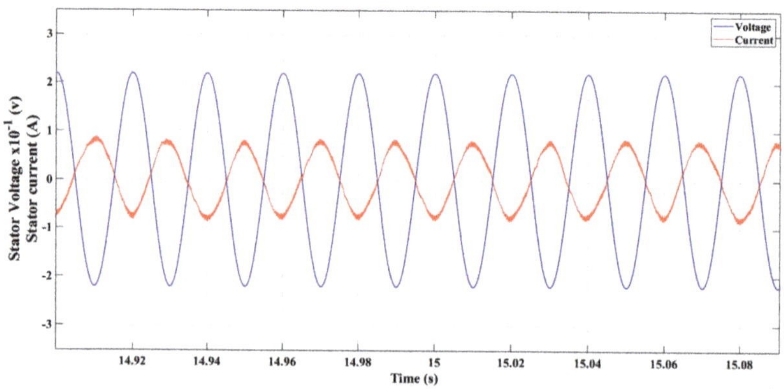

Figure 21. The grid phase voltage and its corresponding phase current waveforms.

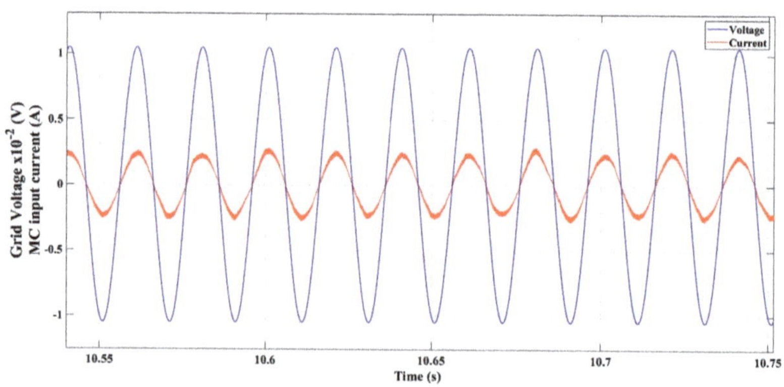

Figure 22. Waveforms of grid phase voltage and MC input current in a sub-synchronous region.

The simulation results confirm that the MC successfully maintained the decoupled active and reactive power control by effectively supplying the necessary rotor-side voltage and frequency, allowing for bidirectional power flow in a range of wind conditions.

Furthermore, a comparison between the suggested ANN-based MPPT and a traditional 2D lookup table-based MPPT, which was developed in [10], was conducted. The tracking performance of the power coefficient (Cp) for both approaches at different wind speeds is shown in Figure 24.

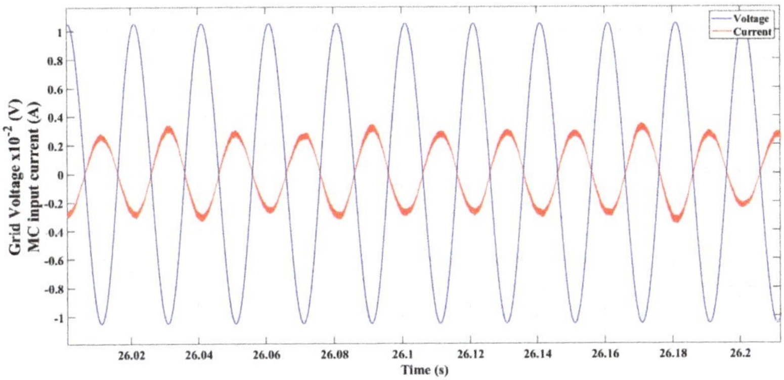

Figure 23. Waveforms of grid phase voltage and MC input current in super-synchronous region.

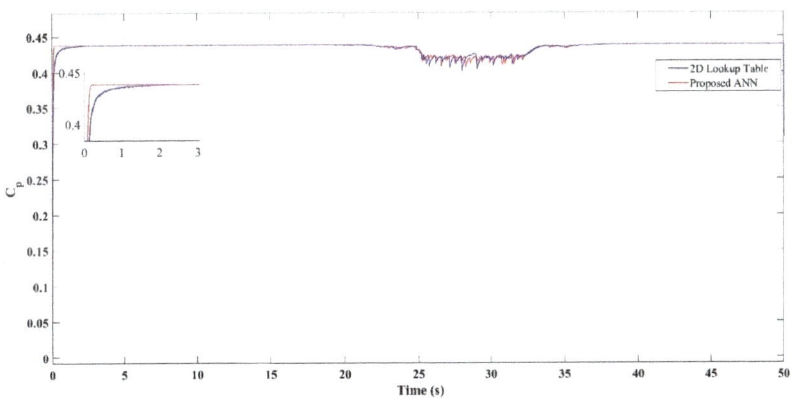

Figure 24. Comparison of power coefficient (Cp) tracking using the proposed ANN-based MPPT and the conventional 2D lookup table method under varying wind speeds.

In comparison to the 2D lookup table technique, the findings show that the ANN-based MPPT achieves faster convergence to the optimal Cp and maintains superior tracking accuracy especially in transient regions. Table 3 provides a summary of a comparison study to further illustrate the benefits of the suggested ANN-based MPPT over the traditional 2D lookup table-based MPPT developed in [10]. This table outlines key differences in adaptability, performance, and implementation aspects between the two approaches, demonstrating the improvements achieved by the proposed method.

Table 3. Comparison between ANN-based and lookup table-based MPPT approaches.

MPPT Method	2D Lookup Table [10]	ANN
Data Requirement	Requires offline-generated lookup table	Trained once; no real-time lookup needed
Adaptability to parameter changes	Limited (static table)	High (ANN can generalize to unseen data)
Convergence speed to C_pmax	Slower	Faster
Control Complexity	Moderate	Slightly higher (due to ANN)
Need for updating data	Requires manual update if parameters change	Self-adaptive after retraining
Memory requirements	Higher (due to table storage)	Lower
Simulation Tool	MATLAB/Simulink	MATLAB/Simulink

8. Conclusions

The design, control, and simulation of the grid-connected WECS are presented in this article. The ANN technique has been used to predict the shaft speed that extracts the maximum possible power from the wind. By utilizing an ANN model, the system dynamically predicts the shaft speed based on the speed of the wind and pitch angle to ensure the extraction of maximum power. The simulation results have illustrated that the usage of the ANN improves adaptability; moreover, it obtains accurate and efficient power tracking under variable wind conditions. Furthermore, it offered faster convergence to the optimal power coefficient and a smoother dynamic response compared to the conventional 2D lookup table method.

There are several benefits using the MC in the rotor side of DFIG as compared to conventional back-to-back converters. Whereby, the need for DC-link capacitors has been removed with the MC, increasing system reliability by lowering the number of elements. It also enables bidirectional power flow, which allows the dynamic transfer of energy between the main grid and rotor of the DFIG in both sub-synchronous and super-synchronous regions. Furthermore, the MC ensures sinusoidal input and output waveforms at both sides, which minimizes harmonic distortion and improves power quality. The simulation results highlight that the combination of ANN-based MPPT together with MC offers a compact, effective, and high-performance solution for DFIG-based WECSs.

Future works might concentrate on putting the proposed control system into real-time hardware. Further investigation will focus on the ANN-based MPPT's scalability for utility-scale turbines, the integration of advanced machine learning models to increase adaptability under rapidly changing situations, and the incorporation of fault detection and diagnostic capabilities to improve system resilience and reliability.

Author Contributions: Conceptualization, M.A.A. and S.S.; formal analysis, S.S.; investigation, M.A.A. and S.S.; methodology, M.A.A.; project administration, S.S.; resources, S.S.; software, M.A.A. and S.S.; supervision, S.S.; validation, M.A.A. and S.S.; writing—original draft, M.A.A.; writing—review and editing, S.S. All authors have read and agreed to the published version of the manuscript.

Funding: This research received no external funding.

Data Availability Statement: The original contributions presented in this study are included in the article. Further inquiries can be directed to the corresponding author.

Conflicts of Interest: The authors declare no conflicts of interest.

References

1. Olabi, A.G.; Obaideen, K.; Abdelkareem, M.A.; AlMallahi, M.N.; Shehata, N.; Alami, A.H.; Mdallal, A.; Hassan, A.A.M.; Sayed, E.T. Wind Energy Contribution to the Sustainable Development Goals: Case Study on London Array. *Sustainability* **2023**, *15*, 4641. [CrossRef]
2. Haq, I.U.; Khan, Q.; Khan, I.; Akmeliawati, R.; Nisar, K.S.; Khan, I. Maximum Power Extraction Strategy for Variable Speed Wind Turbine System via Neuro-Adaptive Generalized Global Sliding Mode Controller. *IEEE Access* **2020**, *8*, 128536–128547. [CrossRef]
3. Bekiroglu, E.; Yazar, M.D. MPPT Control of Grid Connected DFIG at Variable Wind Speed. *Energies* **2022**, *15*, 3146. [CrossRef]
4. Patel, R.; Hafiz, F.; Swain, A.; Ukil, A. Nonlinear Rotor Side Converter Control of DFIG Based Wind Energy System. *Electr. Power Syst. Res.* **2021**, *198*, 107358. [CrossRef]
5. Aljafari, B.; Pamela Stephenraj, J.; Vairavasundaram, I.; Singh Rassiah, R. Steady State Modeling and Performance Analysis of a Wind Turbine-Based Doubly Fed Induction Generator System with Rotor Control. *Energies* **2022**, *15*, 3327. [CrossRef]
6. Boukhezzar, B.; Siguerdidjane, H. Comparison between Linear and Nonlinear Control Strategies for Variable Speed Wind Turbines. *Control Eng. Pract.* **2010**, *18*, 1357–1368. [CrossRef]
7. Barzegar-Kalashani, M.; Seyedmahmoudian, M.; Mekhilef, S.; Stojcevski, A.; Horan, B. Small-Scale Wind Turbine Control in High-Speed Wind Conditions: A Review. *Sustain. Energy Technol. Assess.* **2023**, *60*, 103577. [CrossRef]
8. Le, X.C.; Duong, M.Q.; Le, K.H. Review of the Modern Maximum Power Tracking Algorithms for Permanent Magnet Synchronous Generator of Wind Power Conversion Systems. *Energies* **2022**, *16*, 402. [CrossRef]

9. Pande, J.; Nasikkar, P.; Kotecha, K.; Varadarajan, V. A Review of Maximum Power Point Tracking Algorithms for Wind Energy Conversion Systems. *J. Mar. Sci. Eng.* **2021**, *9*, 1187. [CrossRef]
10. Altun, H.; Sünter, S. Modeling, Simulation and Control of Wind Turbine Driven Doubly-Fed Induction Generator with Matrix Converter on the Rotor Side. *Electr. Eng.* **2013**, *95*, 157–170. [CrossRef]
11. Elumalai, R. Maximum power quality tracking of artificial neural network controller-based double fed induction generator for wind energy conversion system. *Rev. Roum. Sci. Tech.—Ser. Électrotechnique Énergetique* **2024**, *69*, 189–194. [CrossRef]
12. Arnaltes, S.; Rodriguez-Amenedo, J.; Montilla-DJesus, M. Control of Variable Speed Wind Turbines with Doubly Fed Asynchronous Generators for Stand-Alone Applications. *Energies* **2017**, *11*, 26. [CrossRef]
13. Yousefzadeh, M.; Ahmadi Kamarposhti, M. Wind Energy Conversion Systems: A Review on Aerodynamic, Electrical and Control Aspects, Recent Trends, Comparisons and Insights. In *Energy and Environmental Aspects of Emerging Technologies for Smart Grid*; Springer: Cham, Switzerland, 2024; pp. 25–60.
14. Camblong, H.; de Alegria, I.M.; Rodriguez, M.; Abad, G. Experimental Evaluation of Wind Turbines Maximum Power Point Tracking Controllers. *Energy Convers. Manag.* **2006**, *47*, 2846–2858. [CrossRef]
15. Lalouni, S.; Rekioua, D.; Idjdarene, K.; Tounzi, A. Maximum Power Point Tracking Based Hybrid Hill-Climb Search Method Applied to Wind Energy Conversion System. *Electr. Power Compon. Syst.* **2015**, *43*, 1028–1038. [CrossRef]
16. Kumar, S.S.; Jayanthi, K.; Kumar, N.S. Maximum Power Point Tracking for a PMSG Based Variable Speed Wind Energy Conversion System Using Optimal Torque Control. In Proceedings of the 2016 International Conference on Advanced Communication Control and Computing Technologies (ICACCCT), Ramanathapuram, India, 25–27 May 2016; pp. 347–352.
17. Mahfoud, S.; Derouich, A.; Iqbal, A.; El Ouanjli, N. ANT-Colony Optimization-Direct Torque Control for a Doubly Fed Induction Motor: An Experimental Validation. *Energy Rep.* **2022**, *8*, 81–98. [CrossRef]
18. Chetouani, E.; Errami, Y.; Obbadi, A.; Sahnoun, S.; Chojaa, H. Optimization Power Control for Rotor Side Converter of a DFIG Using PSO Evolutionary Algorithm. In *Digital Technologies and Applications, Proceedings of the ICDTA'22, Fez, Morocco, 28–29 January 2022*; Springer: Cham, Switzerland, 2022; pp. 541–551.
19. Guediri, A.; Hettiri, M.; Guediri, A. Modeling of a Wind Power System Using the Genetic Algorithm Based on a Doubly Fed Induction Generator for the Supply of Power to the Electrical Grid. *Processes* **2023**, *11*, 952. [CrossRef]
20. Ali, A.; Irshad, K.; Khan, M.F.; Hossain, M.M.; Al-Duais, I.N.A.; Malik, M.Z. Artificial Intelligence and Bio-Inspired Soft Computing-Based Maximum Power Plant Tracking for a Solar Photovoltaic System under Non-Uniform Solar Irradiance Shading Conditions—A Review. *Sustainability* **2021**, *13*, 10575. [CrossRef]
21. Mohamed, A.Z.; Eskander, M.N.; Ghali, F.A. Fuzzy Logic Control Based Maximum Power Tracking of a Wind Energy System. *Renew. Energy* **2001**, *23*, 235–245. [CrossRef]
22. Sathish Babu, P.; Sundarabalan, C.K.; Balasundar, C.; Santhana Krishnan, T. Fuzzy Logic Based Optimal Tip Speed Ratio MPPT Controller for Grid Connected WECS. *Mater. Today Proc.* **2021**, *45*, 2544–2550. [CrossRef]
23. Ro, K.; Choi, H. Application of Neural Network Controller for Maximum Power Extraction of a Grid-Connected Wind Turbine System. *Electr. Eng.* **2005**, *88*, 45–53. [CrossRef]
24. Tabrizi, Y.H.; Uddin, M.N. Multi-Agent Reinforcement Learning-Based Maximum Power Point Tracking Approach to Fortify PMSG-Based WECSs. *IEEE Trans. Ind. Appl.* **2024**, *60*, 8077–8087. [CrossRef]
25. Chhipa, A.A.; Kumar, V.; Joshi, R.R.; Chakrabarti, P.; Jasinski, M.; Burgio, A.; Leonowicz, Z.; Jasinska, E.; Soni, R.; Chakrabarti, T. Adaptive Neuro-Fuzzy Inference System-Based Maximum Power Tracking Controller for Variable Speed WECS. *Energies* **2021**, *14*, 6275. [CrossRef]
26. Zribi, M.; Alrifai, M.; Rayan, M. Sliding Mode Control of a Variable- Speed Wind Energy Conversion System Using a Squirrel Cage Induction Generator. *Energies* **2017**, *10*, 604. [CrossRef]
27. Mahvash, H.; Taher, S.A.; Rahimi, M.; Shahidehpour, M. Enhancement of DFIG Performance at High Wind Speed Using Fractional Order PI Controller in Pitch Compensation Loop. *Int. J. Electr. Power Energy Syst.* **2019**, *104*, 259–268. [CrossRef]
28. Bossanyi, E.A. The Design of Closed Loop Controllers for Wind Turbines. *Wind Energy* **2000**, *3*, 149–163. [CrossRef]
29. Kalbat, A. Linear Quadratic Gaussian (LQG) Control of Wind Turbines. In Proceedings of the 2013 3rd International Conference on Electric Power and Energy Conversion Systems, Istanbul, Turkey, 2–4 October 2013; pp. 1–5.
30. Daili, Y.; Gaubert, J.-P.; Rahmani, L.; Harrag, A. Quantitative Feedback Theory Design of Robust MPPT Controller for Small Wind Energy Conversion Systems: Design, Analysis and Experimental Study. *Sustain. Energy Technol. Assess.* **2019**, *35*, 308–320. [CrossRef]
31. Rezaei, V. Advanced Control of Wind Turbines: Brief Survey, Categorization, and Challenges. In Proceedings of the 2015 American Control Conference (ACC), Chicago, IL, USA, 1–3 July 2015; pp. 3044–3051.
32. Farrar, N.; Ali, M.H. Cyber-Resilient Converter Control System for Doubly Fed Induction Generator-Based Wind Turbine Generators. *Electronics* **2024**, *13*, 492. [CrossRef]
33. Mahmud, T.; Gao, H. A Comprehensive Review on Matrix-Integrated Single-Stage Isolated MF/HF Converters. *Electronics* **2024**, *13*, 237. [CrossRef]

34. Shapoval, I.; Clare, J.; Chekhet, E. Experimental Study of a Matrix Converter Excited Doubly-Fed Induction Machine in Generation and Motoring. In Proceedings of the 2008 13th International Power Electronics and Motion Control Conference, Poznan, Poland, 1–3 September 2008; pp. 307–312.
35. Cardenas, R.; Pena, R.; Tobar, G.; Clare, J.; Wheeler, P.; Asher, G. Stability Analysis of a Wind Energy Conversion System Based on a Doubly Fed Induction Generator Fed by a Matrix Converter. *IEEE Trans. Ind. Electron.* **2009**, *56*, 4194–4206. [CrossRef]
36. Benbouhenni, H.; Bizon, N. Advanced Direct Vector Control Method for Optimizing the Operation of a Double-Powered Induction Generator-Based Dual-Rotor Wind Turbine System. *Mathematics* **2021**, *9*, 2403. [CrossRef]
37. Abad, G.; López, J.; Rodríguez, M.A.; Marroyo, L.; Iwanski, G. *Doubly Fed Induction Machine*; Wiley: Hoboken, NJ, USA, 2011; ISBN 9780470768655.
38. Sierra-Garcia, J.E.; Santos, M.; Pandit, R. Wind Turbine Pitch Reinforcement Learning Control Improved by PID Regulator and Learning Observer. *Eng. Appl. Artif. Intell.* **2022**, *111*, 104769. [CrossRef]
39. Reyes, V.; Rodriguez, J.J.; Carranza, O.; Ortega, R. Review of Mathematical Models of Both the Power Coefficient and the Torque Coefficient in Wind Turbines. In Proceedings of the 2015 IEEE 24th International Symposium on Industrial Electronics (ISIE), Buzios, Brazil, 3–5 June 2015; pp. 1458–1463.
40. Venturini, M. New Sine Wave in, Sine Wave out Conversion Technique Eliminates Reactive Elements. In Proceedings of the Powercon 7, San Diego, CA, USA, 24–27 March 1980; pp. E3.1–E3.15.
41. Alesina, A.; Venturini, M. Solid-State Power Conversion: A Fourier Analysis Approach to Generalized Transformer Synthesis. *IEEE Trans. Circuits Syst.* **1981**, *28*, 319–330. [CrossRef]
42. Sünter, S. Slip Energy Recovery of a Rotor-Side Field Oriented Controlled Wound Rotor Induction Motor Fed by Matrix Converter. *J. Frankl. Inst.* **2008**, *345*, 419–435. [CrossRef]
43. Narayanaswamy, P.R. Iyer Modelling, Simulation and Real-Time Implementation of a Three-Phase AC to AC Matrix Converter. Ph.D. Thesis, Curtin University, Perth, Australia, 2012.
44. Altun, H.; Sunter, S. Matrix Converter Induction Motor Drive: Modeling, Simulation and Control. *Electr. Eng. (Arch. Elektrotechnik)* **2003**, *86*, 25–33. [CrossRef]

Disclaimer/Publisher's Note: The statements, opinions and data contained in all publications are solely those of the individual author(s) and contributor(s) and not of MDPI and/or the editor(s). MDPI and/or the editor(s) disclaim responsibility for any injury to people or property resulting from any ideas, methods, instructions or products referred to in the content.

energies

MDPI

Article

Safety Assessment of Loop Closing in Active Distribution Networks Based on Probabilistic Power Flow

Wenchao Cai [1,2], Yuan Gao [3], Xiping Zhang [3], Qin Si [1], Jiaoxin Jia [2] and Bingzhen Li [2,*]

1 Inner Mongolia Electric Power Research Institute, Inner Mongolia Electric Power (Group) Co., Ltd., Hohhot 010020, China; buzhidao5477@163.com (W.C.); venus1350053@163.com (Q.S.)
2 Hebei Key Laboratory of Distributed Energy Storage and Microgrid, North China Electric Power University, Baoding 071003, China; jiajx33@163.com
3 Ordos Power Supply Branch, Inner Mongolia Electric Power (Group) Co., Ltd., Ordos 017004, China; gaoyuan1312@163.com (Y.G.); zhangxiping513@163.com (X.Z.)
* Correspondence: 220232213118@ncepu.edu.cn

Abstract: To investigate the security issues of loop-closing operations in medium–low-voltage distribution networks under the influence of stochastic fluctuations from distributed generators (DGs) and loads, probabilistic power flow is introduced for analyzing loop-closing currents in active distribution networks. A novel method combining Latin Hypercube Sampling (LHS) and the Gram–Charlier (GC) series, termed the LHS-GC method, is proposed to calculate the probability distribution of loop-closing currents. By modeling DGs and loads as random variables, their cumulants are efficiently obtained through LHS. Based on a linearized formulation of loop-closing current equations, the cumulants of loop-closing currents are calculated, ultimately reconstructing the probability distribution function of loop-closing currents in active distribution networks. Subsequently, a security assessment framework for loop-closing operations is established using the probability distribution of loop-closing currents. This framework provides a quantitative evaluation from two dimensions: preliminary loop-closing success rate and the severity of current limit violations, offering data-driven decision support for loop-closing operations. Taking the IEEE 34-node distribution network as an example for feeder loop-closing current assessment, the proposed LHS-GC method achieves results with less than 4% deviation from simulation values in terms of cumulative probability distribution of loop-closing currents and safety assessment metrics. Under a sampling scale of 500 points, the computational time is 0.76 s, demonstrating its efficiency and reliability. These outcomes provide actionable references for decision-making support in loop-closing operations of active distribution networks.

Keywords: active distribution network; loop-closing current; probabilistic load flow; cumulant; security assessment

1. Introduction

In recent years, ensuring power-supply reliability has emerged as a critical user demand, driven by economic development. As the power-supply network on the user side, the distribution network in China typically adopts the "closed-loop design, open-loop operation" mode, with a radial network structure [1,2]. When the distribution network lines need to handle faults or undergo maintenance, ring switching can achieve load transfer without interrupting the power supply [3,4]. However, the "connect-first, disconnect-later" strategy modifies the distribution network's topology and power flow. During the loop connection process, steady-state circulating currents and inrush currents may occur, leading to protection malfunctions and line overloads, which in turn threaten the safe and stable operation of the power system [5,6]. Therefore, a method is needed to safely assess the current generated during the loop closure.

In the field of looped connections in classical distribution networks, relevant research has formed a relatively comprehensive theoretical system. In terms of equivalent models for looped connections in distribution networks, reference [7] established a 10 kV distribution network equivalent model considering the equivalent of the main network, feeder load, and transformer. Reference [8] studied the calculation methods for loop current using the superposition theorem and Thevenin's theorem, and analyzed the key factors affecting the magnitude of the current. Reference [9] constructed a typical equivalent network for loops, and designed three methods for calculating loop current in medium- and low-voltage distribution networks, taking into account the complexity of calculation and accuracy.

However, with the widespread adoption of renewable energy, photovoltaic (PV) power generation systems are playing an increasingly critical role in distribution networks. The inherent randomness and volatility of their output power significantly complicate the calculation of loop-closing currents in distribution networks [10,11]. Currently, theoretical research on loop-closing current assessment techniques remains relatively limited, with many studies focusing on power flow optimization and computation in active distribution networks. For instance, reference [12] proposed a nonparametric quasi-Monte Carlo (MC) method based on uniform experimental design for efficient probabilistic power flow calculations. This approach introduces a hybrid discrepancy metric to enhance result accuracy while reducing computational burden. Reference [13] developed a high-order Markov chain-based modeling framework for PV output and load characterization in probabilistic power flow, improving computational efficiency and accuracy through spatio-temporally correlated scenarios generated via joint probability distributions and inverse transform strategies. These studies reveal the potential of integrating probabilistic power flow methodologies into loop-closing current assessment, offering valuable insights for decision support in loop-closing operations. The calculation methods for probabilistic power flow can be categorized into three main classes: the simulation method, analytical method, and approximate method [14]. The simulation method [15,16], represented by the MC method, simulates various uncertain factors through large-scale random sampling. When the sampling is sufficient, the accuracy is high and the applicability is broad, but the computational efficiency is low in complex systems. The approximation method [17,18] directly describes the probabilistic statistical properties of the output random variables, but the computational load increases significantly when solving models with a large number of nodes. The analysis method [19,20] linearizes uncertain variables to obtain the probability distribution of the output random variables, demonstrating excellent performance in computational efficiency.

The cumulant method (CM) is one of the mainstream analytical methods for probabilistic power flow. However, in medium- and low-voltage distribution networks, the probability distributions of many input variables are often unknown. Traditional CM inevitably introduces computational errors due to its reliance on approximated probability distribution functions [21]. To address this limitation, reference [22] improved CM by incorporating maximum entropy-based probability density function approximation, significantly enhancing its accuracy. Building on these advancements, this study focuses on refining the application of CM in loop-closing current assessment. By integrating simulation and analytical approaches, we propose a hybrid method for calculating the probability distribution of loop-closing currents. This approach reduces cumulant calculation errors while maintaining computational efficiency. The technical novelty of the method is summarized as follows:

(1) The Integration of LHS and GC Series: The combination of LHS and GC series for loop-closing current probability distribution calculation in active distribution networks retains the flexibility of simulation methods and the efficiency of analytical approaches.
(2) The Direct Processing of Discrete Sampling Points: LHS is employed to handle discrete sampling points with minimal dependence on predefined input variable distributions. Correlations between sampling points are mitigated through ranking strategies, yielding more accurate cumulants for input variables.
(3) Linear Relationship Between Loop Currents and Nodal Power Injections: By establishing a linear relationship between loop-closing currents and nodal power injections, convolution operations are transformed into algebraic cumulant calculations, significantly enhancing computational efficiency.
(4) Two-Dimensional Safety Assessment Framework: A quantitative framework incorporating preliminary loop-closing success rate and the severity of current limit violations is developed, providing comprehensive and precise quantitative support for operational decision making.
(5) The proposed method is fundamentally based on linearized power flow equations, making it inherently applicable to distribution networks of diverse scales and configurations. Its effectiveness has been validated through a case study on the IEEE 34-node system, demonstrating robust potential for real-world engineering applications.

2. Calculation of Loop Current in Active Distribution Networks

2.1. Calculation of Steady-State Current in a Closed Loop

Taking the typical active distribution network combined-loop operation shown in Figure 1 as an example, the calculation theory and method of combined-loop steady-state current are studied.

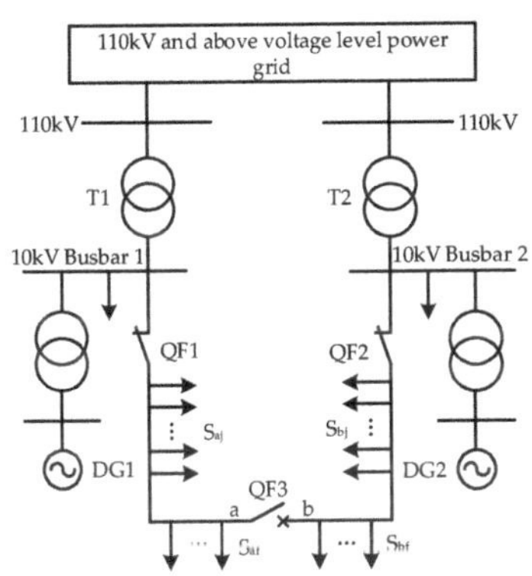

Figure 1. Schematic diagram of active distribution network ring operation.

In the figure, the distribution network operates in an open loop. Two main transformers T1 and T2 are, respectively, configured at 110 kV busbar; the distributed power supplies DG1 and DG2 are, respectively, connected at 10 kV busbar; the outlet of the feeder is controlled by circuit breakers QF1 and QF2, respectively; and the feeder is electrically connected through connection switch QF3. In terms of load distribution, S_{aj} and S_{bj}, respectively, represent the power load of the two feeders, and S_{af} and S_{bf}, respectively, represent the power load of the two sides of the connection switch.

The steady-state current analysis of the ring network is shown in Figure 2. Figure 2a is the network after the closure of the interconnection switch QF3, where $\dot{I}'_a$ and $\dot{I}'_b$ are the steady-state currents of the two feeders after the ring is closed, and $\dot{I}_c$ is the circulating current generated by the closure of the ring; Figure 2b is the network before the ring is closed, where $\dot{I}_a$ and $\dot{I}_b$ are the initial currents of the two feeders before the ring is closed; Figure 2c is the equivalent circuit of the closed-ring network, simplified according to Thevenin's theorem, where the equivalent voltage source $\dot{U}_{oc}$ represents the voltage phasor difference across the interconnection switch before the ring closure, and the equivalent impedance Z_{eq} is the sum of the impedances of the ring network.

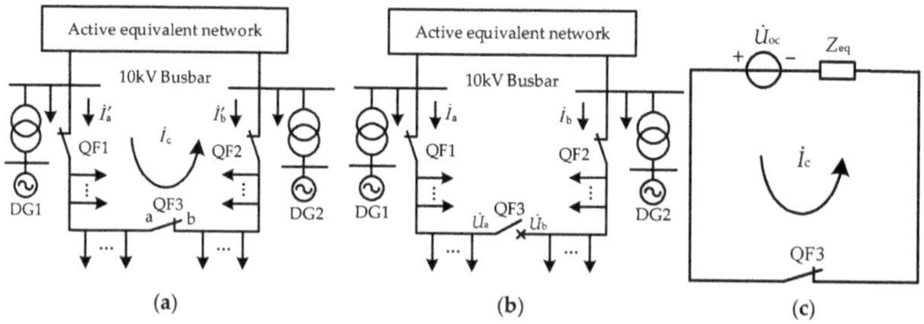

Figure 2. Closed-loop steady-state current analysis diagram. (**a**) Network after ring closure; (**b**) network before closing the loop; (**c**) equivalence network for closed loops.

According to the reference [7], the mesh current in a loop can be represented as:

$$\dot{I}_c = \frac{\dot{U}_{oc}}{\sqrt{3}Z_{eq}} = \frac{\dot{U}_a - \dot{U}_b}{\sqrt{3}Z_{eq}} \tag{1}$$

According to the superposition theorem, when the tie switch QF3 is closed to form a looped network, the steady-state current in the feeders can be decomposed into two independent components: the initial current components $\dot{I}_a'$ and $\dot{I}_b'$ from the original open-loop operation of each feeder, and the circulating current component $\dot{I}_c$ generated by the potential difference at the loop-closing point. By vectorially superimposing these two components, the steady-state current phasors of the feeders under looped operation can be determined as follows:

$$\begin{cases} \dot{I}_a' = \dot{I}_a + \dot{I}_c \\ \dot{I}_b' = \dot{I}_b - \dot{I}_c \end{cases} \tag{2}$$

2.2. Calculation of Loop Impulse Current

The combined-loop operation may cause a large transient current shock, and the influence of the combined-loop shock current on the safety of the power grid needs to be considered. According to reference [10], the loop impulse current instantaneous values i_M and the maximum rms values I_M are:

$$i_M = \sqrt{2}\left(1 + e^{-\frac{0.01}{T_a}}\right) I_c = \sqrt{2} k_m I_c \tag{3}$$

$$I_M = I_c \sqrt{1 + 2(k_m - 1)^2} \tag{4}$$

where I_c is the rms value of the circulating current; T_a is the attenuation time constant of the resultant loop shock current; and $T_a = L_{eq}/R_{eq}$.

As can be seen from Equations (1)–(4), the amplitude of the combined-loop steady-state current and the shock current is determined by the voltage difference on both sides of the break of the link switch, and decreases as the equivalent impedance of the combined-loop network increases. According to Equation (4), the effective values of the maximum shock currents of the feeders at both ends of the combined ring are calculated as follows:

$$\begin{cases} I_{Ma} = I_a + I_c\sqrt{1+2(k_m-1)^2} \\ I_{Mb} = I_b + I_c\sqrt{1+2(k_m-1)^2} \end{cases} \quad (5)$$

Through power flow calculations at the reference node, the voltage magnitudes U_i and phase angles δ_i (i = 1, 2, ..., n) of all nodes in the pre-loop-closing network can be obtained. Let $U = [U_1, \delta_1, \ldots, U_a, \delta_a, U_b, \delta_b, \ldots, U_n, \delta_n]$ denote the state variables of the closed-loop network, and $I = [I'_a, I'_b, I_{Ma}, I_{Mb}]$ represent the effective values of the steady-state currents and inrush currents on both sides of the feeders after loop closing. The loop-closing current equations can then be expressed as:

$$I = G(U) \quad (6)$$

In engineering practice, the inrush coefficient k_m is typically selected within the range of 1.8–1.9 [23,24]. For conservative design purposes, k_m = 1.9 is adopted. Substituting this value into Equation (5), the expanded form of Equation (6) can be derived as:

$$\begin{cases} \dot{I}'_a = \left|\dot{I}_a + \dot{I}_c\right| = g_1(U) \\ \dot{I}'_b = \left|\dot{I}_a - \dot{I}_c\right| = g_2(U) \\ I_{Ma} = I_a + 1.62I_c = g_3(U) \\ I_{Mb} = I_b + 1.62I_c = g_4(U) \end{cases} \quad (7)$$

3. Establishment of System Probability Model

3.1. Power Probability Model of Distributed Power Supply

Distributed household photovoltaic has developed rapidly, and photovoltaic power generation systems will become an important form of renewable energy power generation in low- and medium-voltage distribution networks. The increasing penetration of distributed PV systems, characterized by their randomness and output power variability, has significantly intensified the challenges in performing power flow calculations prior to loop-closing operations. Therefore, not only should the load fluctuation be considered in the calculation of combined loop current, but a probability model for photovoltaic power generation should also be constructed. The illumination intensity r can be approximately regarded as a Beta distribution over a finite period of time, and its probability density function is [13]:

$$f(r) = \frac{\Gamma(\alpha+\beta)}{\Gamma(\alpha)\Gamma(\beta)}\left(\frac{r}{r_{max}}\right)^{\alpha-1}\left(1-\frac{r}{r_{max}}\right)^{\beta-1} \quad (8)$$

where r_{max} is the maximum illumination intensity, α and β are the two shape parameters of the Beta distribution, and Γ is the Gamma function.

In practice, the shape parameters of the Beta distribution can be determined by statistical analysis, and two shape parameters in the Beta distribution are calculated from the expectations μ and variance σ of the given time-series data [11]:

$$\begin{cases} \alpha = \mu\left[\frac{\mu(1-\mu)}{\sigma^2} - 1\right] \\ \beta = (1-\mu)\left[\frac{\mu(1-\mu)}{\sigma^2} - 1\right] \end{cases} \quad (9)$$

The power probability characteristic of the photovoltaic power supply satisfies the following requirements:

$$P_V(r) = r \cdot A \cdot \eta \quad (10)$$

where P_V is the total output power; A is the total area of the photovoltaic panel, and η is the photoelectric conversion efficiency.

It can be seen from Equation (10) that the output power of the photovoltaic system has a positive correlation with the illumination. In combination with Equation (8), the probability density function of the light intensity and the photovoltaic output force can be derived:

$$f(P_V) = \frac{\Gamma(\alpha+\beta)}{\Gamma(\alpha)\Gamma(\beta)}(\frac{P}{V_{max}})^{\alpha-1}(1-\frac{P}{V_{max}})^{\beta-1} \tag{11}$$

where P_{Vmax} is the maximum power of photovoltaic power generation, which meets $P_{Vmax} = r_{max}A\eta$.

In practice, through the regulation of the control system, photovoltaic power generation units can achieve stable control of the power factor, so they can be modeled as PQ nodes in the load flow calculation of the distribution network. Thus, under the known power factor angle of φ, the reactive output power of the photovoltaic array can be obtained as Q_V [14].

$$Q_V = P_V \tan\varphi \tag{12}$$

After applying PQ node equivalent processing to the photovoltaic power source, it can be regarded as a load to derive the Thevenin equivalent impedance expression Z_{Veq}:

$$Z_{veq} = -\frac{P_V U_V^2}{P_V^2+Q_V^2} - j\frac{Q_V U_V^2}{P_V^2+Q_V^2} \tag{13}$$

3.2. Load Power Probability Model

Using a normal distribution to approximate the uncertainty of reactive power load, assuming the corresponding parameters for active power and reactive power are μ_P, σ_P and μ_Q, σ_Q, the probability density functions of active power P_L and reactive power Q_L can be expressed as [20]:

$$\begin{cases} f(P_L) = \frac{1}{\sqrt{2\pi}\sigma_P}\exp\left[-\frac{(P_L-\mu_P)}{2\sigma_P^2}\right] \\ f(Q_L) = \frac{1}{\sqrt{2\pi}\sigma_Q}\exp\left[-\frac{(Q_L-\mu_Q)}{2\sigma_Q^2}\right] \end{cases} \tag{14}$$

4. Solution of Cumulants for Discrete Input Variables

4.1. Latin Hypercube Sampling

When calculating the joint probability of the load considering the randomness of photovoltaic power generation, it is necessary to establish a method based on stochastic power flow calculations. LHS can effectively reflect the overall distribution of random variables through sampling values. This method divides the range of each variable equally and samples uniformly within each interval, thereby improving the accuracy of the simulation. Therefore, compared to MC sampling, it can obtain more accurate sample mean and variance values with fewer sampling numbers.

LHS mainly has two steps of sampling and arrangement. The sampling step ensures that the sample distribution area can be sufficiently collected, assuming $X_1, X_2, \ldots$, and X_p are p input random variables, where the cumulative probability distribution function of X_p is:

$$Y_p = F_p(X_p) \tag{15}$$

The sampling scale is N, the value-taking interval (0,1) of Yp is divided into N intervals, and the length of each interval is $1/N$. By setting the sample point Y_p at the center of each interval, the i-th sample value of the corresponding input random variable X_p can be obtained from the inverse function:

$$X_{pi} = F_p^{-1}(\frac{i-0.5}{N}) \tag{16}$$

The sampling matrix of each input random variable is $1 \times N$, and all p input random variables are processed to form an initial sampling matrix of $p \times N$, so that the sampling step is finished, and the next order is carried out. The sorting step randomly arranges the position of each element in each row, reduces the correlation between each random variable of the initial sampling matrix, and obtains the final sampling matrix. Calculations of the order of origin moments $\alpha_p^{(v)}$ and cumulants $\gamma_p^{(v)}$ of the input random variable X_p are then performed based on the final sampling matrix.

4.2. Latin Hypercube Sampling for Discrete Data

The LHS requires the cumulative probability distribution function $Y_p = F_p(X_p)$ of the random variable X_p as input. However, in engineering applications, many of the operating parameters and historical data of variables are often discrete, making it impossible to directly obtain the cumulative probability distribution function of these discrete data. Therefore, a method is needed to determine semi-variates using LHS based on discrete data.

For a subsample set of the random variable X_p, when it contains N discrete observations, it can be represented as $\{x_{p1}, x_{p2}, x_{p3}, \ldots, x_{pN}\}$. Sort the subsample set of each input random variable in ascending order and calculate the empirical distribution function of X_p:

$$H_p(X_p) = \begin{cases} 0,\ x_p < x'_{p1} \\ \frac{k}{n}, x'_{pk} \le x_p \le x'_{p(k+1)}, k = 1,2,\ldots,N-1 \\ 1, x_p \ge x'_{p(N+1)} \end{cases} \tag{17}$$

According to the empirical distribution function $H_p(X_p)$, a sample set of X_p $\{x''_{p1}, x''_{p2}, x''_{p3}, \ldots, x''_{pN}\}$ is obtained using LHS, and the moments of different orders of X_p are calculated, respectively:

$$\alpha_p^{(v)} = \frac{1}{N}\sum_{i=1}^{N} x_{pk}^{(v)}, k = 1,2,\ldots,n \tag{18}$$

Then, based on the relationship between cumulants and the moment about the origin, we obtain the cumulants of each order for X_p.

5. Probabilistic Calculation of Loop-Closing Currents and Line Losses

5.1. Linearization of the Loop Current Equation

Let $W = [P_1, Q_1, \ldots, P_n, Q_n]$ be the injection power of each node in the ring network, and the matrix form of the node power equation is:

$$W = F(U) \tag{19}$$

The state variable U of the closed-loop network system can be decomposed into the expected value U_0 and the disturbance amount ΔU, which satisfies the relationship $W_0 = F(U_0)$ with the expected value of the node injection power W_0. Expanding Equation (19) according to the Taylor series and ignoring higher-order terms yields:

$$\Delta W = J_0 \Delta U \tag{20}$$

where J_0 is the Jacobian matrix, defined as:

$$J_0 = \left.\frac{\partial F(U)}{\partial U}\right|_{U = U_0} \tag{21}$$

Similarly, the loop current equation can also be subjected to similar linearization processing, resulting in:

$$\Delta I = G_0 \Delta U = G_0 J_0^{-1} \Delta W \tag{22}$$

where G_0 is the coefficient matrix, defined as:

$$G_0 = \left.\frac{\partial G(U)}{\partial U}\right|_{U = U_0} \tag{23}$$

Equation (22) shows that the combined-loop current can be expressed as a linear combination of the variable ΔW, providing a theoretical basis for obtaining the probability distribution of the combined-loop current based on cumulants.

5.2. Loop-closing Current Probability Density Function

Apply the LHS method to process discrete data to obtain the cumulants $\Delta W_{\mathrm{P}}^{(\mathrm{k})}$ and $\Delta W_{\mathrm{L}}^{(\mathrm{k})}$ for photovoltaic power generation and load. Due to the additivity of cumulants, by summing the cumulants of each node's load power $\Delta W_{\mathrm{L}}^{(\mathrm{k})}$ and the cumulant of the photovoltaic output $\Delta W_{\mathrm{P}}^{(\mathrm{k})}$, we obtain the cumulative amount of node injection power $\Delta_{W}^{(\mathrm{k})}$.

$$\Delta_{\mathrm{W}}^{(\mathrm{k})} = \Delta W_{\mathrm{L}}^{(\mathrm{k})} + \Delta W_{\mathrm{P}}^{(\mathrm{k})} \tag{24}$$

Refer to the linear properties of the cumulant and combine with Equation (22) to obtain the cumulant of the combined ring current $\Delta I^{(k)}$.

$$\Delta I^{(k)} = \left(G_0 J_0^{-1}\right)^k \Delta W^{(k)} \tag{25}$$

After obtaining the cumulants of each order of the circular current, the cumulative probability distribution function $F(x)$ can be constructed using the Gram–Charlier series expansion method, as follows [22]:

$$F(x) = \Psi(x) + \frac{c_1}{1!}\Psi'(x) + \frac{c_2}{2!}\Psi''(x) + \ldots \tag{26}$$

where $\Psi(x)$ is the cumulative distribution function of the standard normal distribution; the coefficients c_i can be obtained using the semi-invariants of each order.

5.3. Probability Function of Power Loss in Distribution Networks

Power loss is a critical comprehensive indicator for evaluating the economic and technical performance of power grid operation. The integration of DGs into distribution networks significantly complicates power loss calculations. To reduce computational complexity, power loss can be defined as the difference between injected power and output power. The cumulants of different orders for line losses in the distribution network can be calculated as follows:

$$\Delta P_{\mathrm{loss}}^{(k)} = P_{\mathrm{s}}^{(k)} + \sum_{i=1}^{g} P_{i\_\mathrm{DG}}^{(k)} - \sum_{j=1}^{h} P_{j\_\mathrm{L}}^{(k)} \tag{27}$$

where $\Delta P_{\text{loss}}^{(k)}$ is the k-th order cumulant of distribution network power loss; $P_{\text{s}}^{(k)}$ is the k-th order cumulant of the output power from the main power source; $\sum_{i=1}^{g} P_{i\_\text{DG}}^{(k)}$ is the sum of k-th order cumulants from g distributed generators; $\sum_{j=1}^{h} P_{j\_\text{L}}^{(k)}$ is the sum of k-th order cumulants from h loads.

6. Safety Assessment Indicators for Distribution Network Ring Connection

6.1. Initial Success Rate of Loop Closure

The conditions for the safe loop operation of the distribution network are (1) the steady-state current of each feeder after looping does not exceed its maximum carrying capacity; (2) the impact current of each feeder during the closing loop is below the fast trip action threshold of the protective device.

Let $F_1(x)$, $F_2(x)$, $F_3(x)$, and $F_4(x)$ be the cumulative distribution functions of the combined-loop currents I'_{a}, I'_{b}, I_{Ma}, and I_{Mb}, respectively. The current carrying limits of the two feeder lines are $I_{\text{max.a}}$ and $I_{\text{max.b}}$, and the settings for the current instantaneous protection are $I_{\text{setI;.a}}$ and $I_{\text{setI.b}}$. Then, the probability of the combined-loop current exceeding the safety threshold can be expressed as [25]:

$$\begin{cases} P_1 = P(\dot{I}'_{\text{a}} \geq I_{\text{max.a}}) = 1 - F_1(I_{\text{max.a}}) \\ P_2 = P(\dot{I}'_{\text{a}} \geq I_{\text{max.b}}) = 1 - F_2(I_{\text{max.b}}) \\ P_3 = P(\dot{I}_{\text{Ma}} \geq I_{\text{set.a}}) = 1 - F_3(I_{\text{set.a}}) \\ P_4 = P(\dot{I}_{\text{Mb}} \geq I_{\text{set.b}}) = 1 - F_4(I_{\text{set.b}}) \end{cases} \tag{28}$$

The preliminary success rate of grid connection is quantitatively described to assess the impact of distributed power sources on exceeding the grid connection current limit, providing data support for grid connection decisions. The expression for the preliminary success rate P is as follows:

$$P = (1 - P_1)(1 - P_2)(1 - P_3)(1 - P_4) \tag{29}$$

If P is greater than 95%, this operation is defined as a preliminary successful closure.

6.2. Excessive Ring Current Limit

Although the initial success rate of the loop closure can indicate a high probability of meeting the requirements for loop-closure current, certain loop-closure currents with lower probabilities may cause serious violation issues. Therefore, it is necessary to define the maximum violation rate of the loop-closure current η_{M} and the average violation rate η_{A} as follows [25]:

$$\eta_{\text{M}} = \left(\frac{I_{\text{F}}}{I_{\text{Lmax}}} - 1 \right) \times 100\% \tag{30}$$

$$\eta_{\text{A}} = \left(\frac{\int_{I_{\text{Lmax}}}^{I_{\text{F}}} x \cdot f(x)\text{d}x}{\int_{I_{\text{Lmax}}}^{I_{\text{F}}} f(x)\text{d}x} - 1 \right) \times 100\% \tag{31}$$

where I_{F} is the current value corresponding to the cumulative probability distribution of the line current at 99.9%, I_{Lmax} is the maximum current-carrying capacity of the line, and $f(x)$ is the probability density function of the line current.

Since the power system has a certain overload capacity, it is considered that the safe operation of the distribution network requires that the η_{M} does not exceed 10% and η_{A}

does not exceed 5%. If a preliminary loop closing is unsuccessful but the exceeding current rate meets the requirements, then the part of the current that does not meet the preliminary loop-closing success rate is determined to be the safe current. In summary, using preliminary loop-closing success rate, maximum exceeding rate, and average exceeding rate as quantitative standards for assessing the safety of loop-closing operations provides data references for the operators' loop-closing decisions.

In summary, the process of the LHS-GC method based on LHS and Gram–Charlier series is shown in Figure 3. The arrows in Figure 3 indicate the next step.

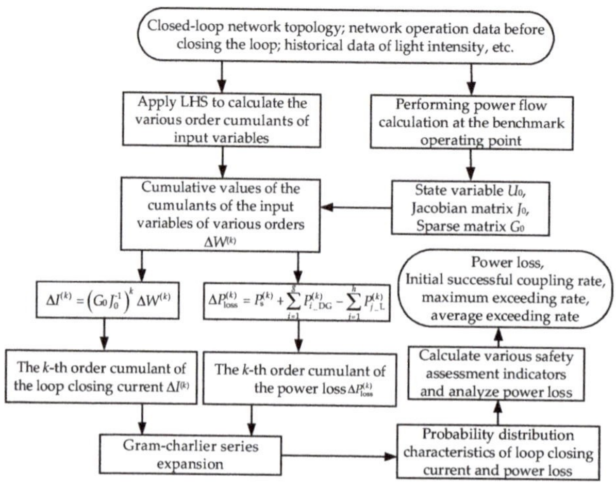

Figure 3. LHS-GC calculation process.

7. Example Verification

This paper uses the IEEE 34-node distribution network system as an example to verify the proposed method. The standard system is simplified to a single voltage level by removing transformers and voltage regulators from the lines. Due to the text length constraints in the main body, the line impedance data of IEEE 34 nodes and the normal distribution data of loads are presented in Appendices A and B. Based on the simplified distribution network, photovoltaic power sources are configured using solar radiation intensity samples from a region in South China as input data. The system topology and an example of the tie line are shown in Figure 4. In Figure 4, the black solid line represents the existing distribution line, and the blue dashed line represents the tie line that has not been closed.

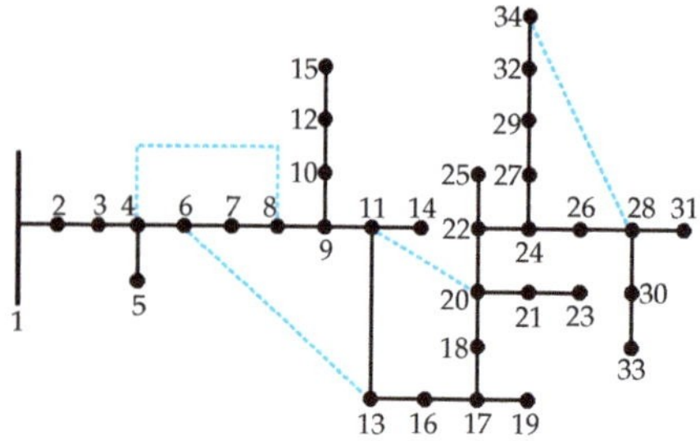

Figure 4. A schematic diagram of the IEEE 34-node distribution network system.

Based on the sample data of solar radiation intensity within a day, the shape parameters of the Beta distribution are calculated to be $\alpha = 0.679$ and $\beta = 1.778$, with a maximum solar radiation intensity of 1.134 kW/m^2, as shown in Figure 5. Take the reference capacity S_B as 1 MVA and the reference voltage U_B as 24.9 kV.

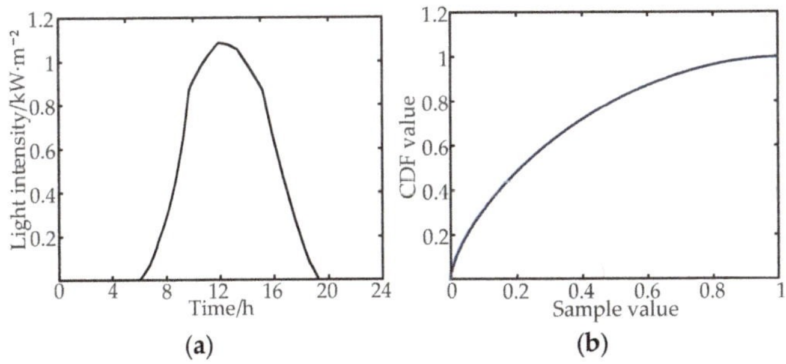

Figure 5. A photovoltaic sample curve for a region in South China. (**a**) Diurnal variation curve of light intensity; (**b**) light intensity distribution characteristic curve.

Based on variations in PV integration nodes, loop-closing line locations, and load standard deviations, two cases are defined as shown in Table 1. These cases are selected to analyze the safety of loop-closing operations and power losses.

Table 1. Configuration of loop-closing operating conditions.

Case	PV Integration Nodes	Loop-Closing Line Locations	Load Standard Deviation
1	23, 29	28–34	30%
2	6, 23	6–13	20%

7.1. Probabilistic Calculation of Loop-Closing Currents and Power Losses

This case study uses loop-closing currents obtained from MATLAB/SIMULINK simulations as a reference to compare the results of three methods: MC method, traditional cumulant method, and the proposed LHS-GC method, with a sampling size of 500 for both MC and LHS. The cumulative probability distributions of loop-closing currents for Case 1 and Case 2 are illustrated in Figures 6 and 7, respectively.

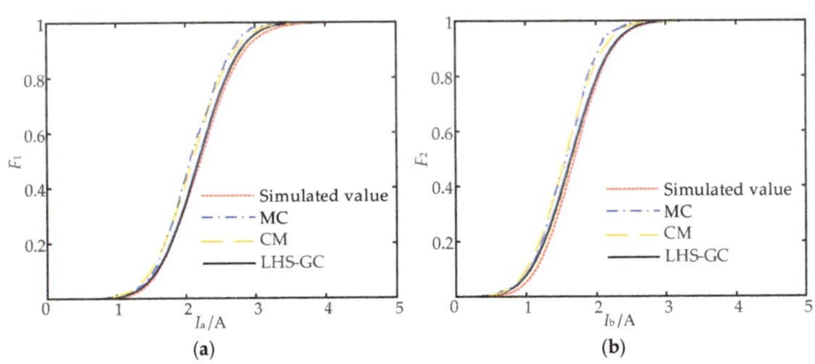

Figure 6. *Cont.*

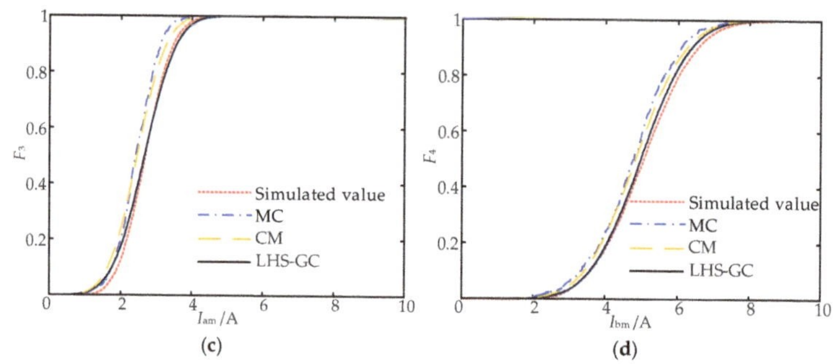

Figure 6. Cumulative probability distribution of loop-closing currents for Case 1. (**a**) Line 26–28 loop steady-state current; (**b**) line 32–34 loop steady-state current; (**c**) line 26–28 loop impulse current; (**d**) line 32–34 loop impulse current.

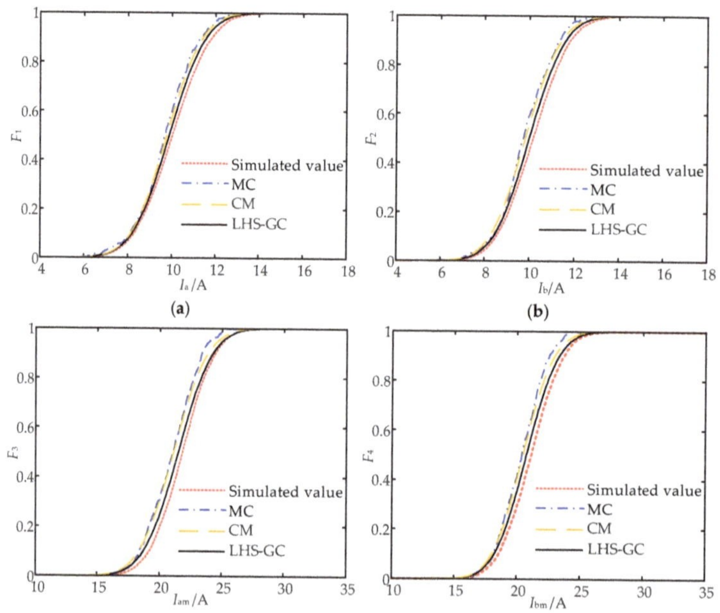

Figure 7. Cumulative probability distribution of loop-closing currents for Case 1. (**a**) Line 4–6 loop steady-state current; (**b**) line 11–13 loop steady-state current; (**c**) line 4–6 loop impulse current; (**d**) line 11–13 loop impulse current.

As shown in Figures 6 and 7, compared to the MC method and traditional CM, the proposed LHS-GC method achieves higher accuracy, with its cumulative probability distribution curves aligning more closely with simulation results. Due to limitations in sampling scale, the traditional CM demonstrates higher precision than the MC method under the current configuration.

Figure 8 illustrates the deviations of the three methods from simulation values at the 90% cumulative probability level. Taking Case 1 as an example, the calculation errors of the LHS-GC method for steady-state currents and inrush currents in the two feeders are 3.49%, 6.33%, 3.4%, and 3.11% lower than those of the MC method, and 1.72%, 3.1%, 3.09%, and 1.8% lower than those of the CM, respectively. Furthermore, the current calculation errors of the LHS-GC method are consistently below 3%, confirming its superior performance in approximating the cumulative probability distribution function of feeder loop-closing currents. These results validate the reliability of the LHS-GC method for loop-closing safety assessments.

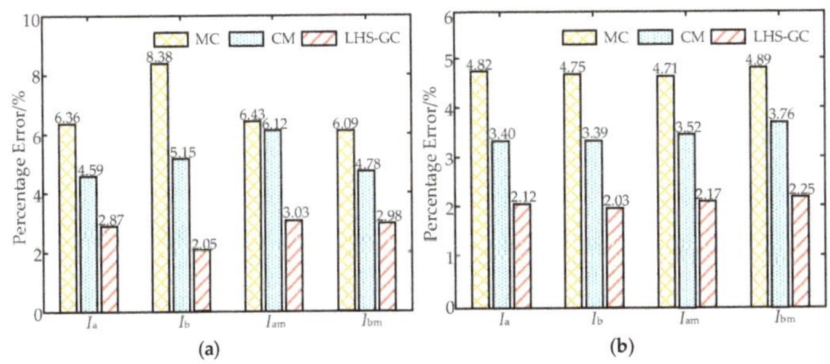

Figure 8. Feeder current error at 90% cumulative distribution probability. (**a**) Case 1; (**b**) Case 2.

Table 2 compares the computation times of the methods. The LHS-GC method requires shorter single-cycle time and total duration, achieving a 68% speed improvement compared to the MC method. Under identical sampling scales, it exhibits significant speed advantages. Therefore, compared to the traditional CM, the LHS-GC method achieves higher accuracy with only a minor sacrifice in computational efficiency, making it a practical tool for real-time operator decision making.

Table 2. Comparison of computation time.

Computational Methods	Single Cycle Time/s	Total Duration/s
MC method	0.0019	2.43
CM	-	0.15
LHS-GC method	0.0001	0.76

Under both operating cases, the power losses of the entire network before and after loop closing are shown in Figure 9. For Case 1, the power losses showed no significant variation between pre- and post-loop-closing operations. In contrast, for Case 2, the power losses decreased after loop closing compared to pre-loop-closing levels, contributing to the economic operation of the distribution network. The proposed LHS-GC method accurately captures the variations in power losses, exhibiting errors within 8% compared to simulation values at the 90% cumulative probability level. This demonstrates its capability to provide a reliable reference for assessing the economic and technical performance of post-loop-closing distribution networks.

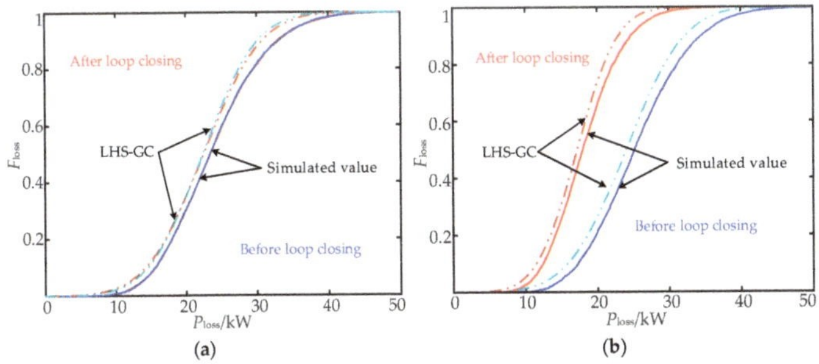

Figure 9. Power loss in distribution network before and after loop closure. (**a**) Case 1; (**b**) Case 2.

7.2. Loop-Closing Safety Assessment

Safety assessments for loop-closing operations are conducted using the cumulative probability distribution curves of feeder loop-closing currents calculated by the LHS-GC method. Based on the 3-sigma principle of load normal distribution, the maximum allowable current-carrying capacity of branches is determined. The instantaneous current protection settings are derived from the maximum short-circuit current at the end of the protected line during a fault. The calculated results for the maximum allowable current-carrying capacity and protection settings under Case 1 and Case 2 are summarized in Table 3.

Table 3. The current indicator values of Case 1 and Case 2.

Feeder	Maximum Allowable Current-Carrying Capacity/A	Instantaneous Current Protection Settings/A
26–28	3.24	13.21
32–34	2.62	36.50
4–6	12.93	143.94
11–13	12.73	100.68

The safety assessment results of loop-closing currents are presented in Tables 4 and 5. In Case 1, the proposed LHS-GC method exhibits errors of 0.69% and 1.17% in the maximum violation rates for feeders 26–28 and 32–34 compared to simulation values, respectively. The errors in average exceedance rates are 0.36% and 0.95%, respectively. In Case 2, the errors of maximum violation rates and average exceedance rates calculated by LHS-GC for feeders 4–6 and 11–13 are 1.53%, 0.59%, and 0.86%, 0.67%, respectively, all within 2%. Additionally, the preliminary loop-closing success rates calculated by LHS-GC for both cases closely align with simulation values, exhibiting errors of 1.26% and 1.81%.

Table 4. Assessment indices of loop-closing safety under Case 1.

<table>
<tr><th rowspan="2">Loop Current</th><th colspan="2">Preliminary Closure Probability</th><th colspan="2">Maximum Overrun Rate</th><th colspan="2">Average Exceedance Rate</th></tr>
<tr><th>Simulated Value</th><th>LHS-GC</th><th>Simulated Value</th><th>LHS-GC</th><th>Simulated Value</th><th>LHS-GC</th></tr>
<tr><td>26–28 Steady-state current</td><td rowspan="4">94.66%</td><td rowspan="4">95.92%</td><td>12.43%</td><td>13.12%</td><td>4.37%</td><td>4.01%</td></tr>
<tr><td>32–34 Steady-state current</td><td>13.29%</td><td>14.46%</td><td>4.89%</td><td>3.94%</td></tr>
<tr><td>26–28 impact current</td><td>—</td><td>—</td><td>—</td><td>—</td></tr>
<tr><td>32–34 impact current</td><td>—</td><td>—</td><td>—</td><td>—</td></tr>
</table>

In summary, an analysis of these two cases demonstrates that the LHS-GC method can effectively assess the safety of loop-closing operations in active distribution networks, even with limited samples. The entire process—from calculating the cumulative probability distribution of loop-closing currents to deriving safety assessment metrics—requires only 0.76 s, meeting the timeliness requirements for engineering applications.

Table 5. Assessment indices of loop-closing safety under Case 2.

Loop Current	Preliminary Closure Probability		Maximum Overrun Rate		Average Exceedance Rate	
	Simulated Value	LHS-GC	Simulated Value	LHS-GC	Simulated Value	LHS-GC
4–6 Steady-state current	91.35%	93.16%	15.33%	16.86%	5.11%	5.97%
11–13 Steady-state current			14.72%	15.31%	5.25%	5.92%
4–6 impact current			—	—	—	—
11–13 impact current			—	—	—	—

8. Conclusions

To address the challenges in the safety assessment of loop-closing operations for active distribution networks, this paper proposes an LHS-GC-based framework for probabilistic loop-closing current analysis and safety evaluation. Validation on the IEEE 34-node system yields the following conclusions:

(1) Compared to the MC method, the LHS-GC method significantly improves both accuracy and computational speed under identical sampling scales. Compared to the traditional CM, it achieves higher accuracy while maintaining comparable speed. With a sampling size of 500, the method completes calculations in 0.76 s with probability distribution errors below 4%.

(2) Under the two-dimensional safety assessment framework (preliminary success rate and severity of current violations), the LHS-GC method achieves errors within 3% for all metrics. The total computation time of 0.76 s fulfills real-time operational demands for active distribution networks.

(3) The LHS-GC method accurately captures the probabilistic distribution of loop-closing currents and safety metrics across diverse scenarios, including varying PV integration nodes, loop-closing line locations, and load standard deviations. Rooted in linearized power flow equations, the method is inherently adaptable and holds potential for application in distribution networks with various configurations.

Author Contributions: Conceptualization, W.C.; methodology, Y.G.; formal analysis, X.Z.; data curation, Q.S.; writing—original draft preparation, B.L.; writing—review and editing, W.C.; funding acquisition, J.J. All authors have read and agreed to the published version of the manuscript.

Funding: Science and technology project funding of Inner Mongolia Power Group: 2024-4-47; Supported by the National Natural Science Foundation of China: 52207102.

Data Availability Statement: Data are unavailable due to privacy restrictions.

Conflicts of Interest: Authors Wenchao Cai, Qin Si, Yuan Gao and Xiping Zhang were employed by the company Inner Mongolia Electric Power (Group) Co., Ltd. The remaining authors declare that the research was conducted in the absence of any commercial or financial relationships that could be construed as a potential conflict of interest.

Abbreviations

The following abbreviations are used in this manuscript:

PV	Photovoltaic
MC	Monte Carlo method
CM	Cumulant method
LHS	Latin hypercube sampling
GC	Gram–Charlier
DG	Distributed generation
QF	Circuit breaker
S_{aj}, S_{bj}	the power load of the two feeders
S_{af}, S_{bf}	the power load of the two sides of the connection switch
$\dot{I}_a, \dot{I}_b$	the initial currents of the two feeders
$\dot{I}_a', \dot{I}_b'$	Current of the feeders after closing the loop
$\dot{I}_c$	the circulating current
$\dot{U}_{oc}$	Voltage difference across the contact switch
Z_{eq}	Equivalent impedance of the loop
i_M	Instantaneous values of the loop impulse current
I_M	Maximum effective value of loop impulse current
I_c	the rms value of the circulating current
T_a	Decay time constant of the loop impulse current
R_{eq}	Equivalent reactance of the loop
L_{eq}	Equivalent resistance of the loop
I_{Ma}, I_{Mb}	Maximum impulse current effective value of the two feeders
k_m	Impact coefficient
r	Illumination intensity
α, β	Shape parameters for beta distribution
Γ	Gamma function
μ	Expectations for data in a given time series
σ	Variance of data for a given time series
P_V, Q_V	Photovoltaic array output power
A	Total area of photovoltaic panels
η	Photoelectric conversion efficiency
P_{Vmax}	Maximum power for photovoltaic power generation
X_p	Input random variables
Y_p	Cumulative distribution function of X_p
$\alpha_p^{(v)}$	Moment matrix of each order
$\gamma_p^{(v)}$	Partial invariants of each order
$H_p(X_p)$	Empirical distribution function
W	Injected power of each node in the ring network
J_0	Jacobian matrix
$\Delta W_P^{(k)}$	Cumulants of photovoltaic power at various stages
$\Delta W_L^{(k)}$	Cumulants of load at each level
$\Delta W^{(k)}$	Cumulants of the node injection power of various orders
$\Delta I^{(k)}$	Cumulants of the loop current
$\Psi(x)$	the cumulative distribution function of the standard normal distribution
$I_{max.a}, I_{max.b}$	the current carrying limits of the two feeder lines
$I_{setI.a}, I_{setI.b}$	the settings for the current instantaneous protection
P	The probability of the closed-loop current exceeding the safety threshold
η_M	the maximum violation rate of loop-closure current
η_A	the average violation rate
S_B	Benchmark capacity of distribution network
U_B	Reference voltage of the distribution network

Appendix A

Table A1. Original data table of IEEE 34-node line.

Starting Node	Leaf Node	Resistor (p.u.)	Reactance (p.u.)	Electricity (p.u.)
1	2	2.030825×10^{-3}	8.953083×10^{-4}	$4.500000\text{E} \times 10^{-7}$
2	3	1.361754×10^{-3}	6.003423×10^{-4}	3.000000×10^{-7}
3	4	2.536956×10^{-2}	1.118441×10^{-2}	5.500000×10^{-6}
4	5	4.596905×10^{-3}	2.026589×10^{-3}	1.000000×10^{-6}
4	6	2.951780×10^{-2}	1.301320×10^{-2}	6.500000×10^{-6}
6	7	2.340171×10^{-2}	1.031687×10^{-2}	5.000000×10^{-6}
7	8	7.871413×10^{-4}	3.470187×10^{-4}	0.000000
8	9	2.440138×10^{-4}	1.075758×10^{-4}	0.000000
9	10	1.346012×10^{-3}	5.934020×10^{-4}	2.500000×10^{-7}
9	11	8.036713×10^{-3}	3.543061×10^{-3}	1.500000×10^{-6}
10	12	3.790085×10^{-2}	1.670895×10^{-2}	8.000000×10^{-6}
11	13	6.611987×10^{-4}	2.914957×10^{-4}	1.000000×10^{-7}
11	14	2.385038×10^{-3}	1.051467×10^{-3}	5.000000×10^{-8}
12	15	1.081532×10^{-2}	4.768037×10^{-3}	2.350000×10^{-5}
13	16	1.608917×10^{-2}	7.093062×10^{-3}	3.550000×10^{-5}
14	17	4.093135×10^{-4}	1.804497×10^{-4}	0.000000
17	18	2.899042×10^{-2}	1.278070×10^{-2}	6.000000×10^{-6}
17	19	1.836401×10^{-2}	8.095946×10^{-3}	4.045000×10^{-5}
18	20	7.871413×10^{-4}	3.470187×10^{-4}	0.000000
20	21	3.064467×10^{-3}	6.580539×10^{-3}	3.000000×10^{-6}
20	22	3.856993×10^{-3}	1.700392×10^{-3}	5.000000×10^{-7}
21	23	8.312212×10^{-3}	3.664518×10^{-3}	$1.500000\text{E} \times 10^{-6}$
22	24	4.589034×10^{-3}	2.023119×10^{-3}	1.000000×10^{-6}
22	25	1.275169×10^{-3}	5.621703×10^{-4}	2.500000×10^{-7}
24	26	1.590025×10^{-3}	7.009778×10^{-4}	5.000000×10^{-8}
24	27	2.203996×10^{-4}	9.716525×10^{-5}	5.000000×10^{-9}
26	28	2.109539×10^{-3}	9.300102×10^{-4}	5.000000×10^{-7}
27	29	1.062641×10^{-3}	4.684753×10^{-4}	2.000000×10^{-7}
28	30	2.203996×10^{-4}	9.716525×10^{-5}	5.000000×10^{-9}
28	31	6.769415×10^{-4}	2.984361×10^{-4}	1.500000×10^{-7}
29	32	2.865194×10^{-3}	1.263148×10^{-3}	5.000000×10^{-7}
30	33	2.558431×10^{-3}	1.684948×10^{-3}	5.000000×10^{-7}
32	34	4.171849×10^{-4}	1.839199×10^{-4}	5.000000×10^{-8}

Appendix B

Table A2. Load data table of IEEE 34 nodes (for example, a standard deviation of 30%).

Node	Active Power Load		Reactive Power Load	
	Expectation	Variance	Expectation	Variance
2	1.910000×10^{-2}	5.730000×10^{-3}	9.870000×10^{-3}	2.960000×10^{-3}
4	5.290000×10^{-3}	1.587000×10^{-3}	2.740000×10^{-3}	8.220000×10^{-4}
9	1.300000×10^{-4}	3.900000×10^{-5}	7.000000×10^{-5}	2.100000×10^{-5}
10	4.596905×10^{-3}	3.390000×10^{-3}	5.840000×10^{-3}	1.752000×10^{-3}
11	1.490000×10^{-2}	4.470000×10^{-3}	7.710000×10^{-3}	2.313000×10^{-3}
12	1.184000×10^{-2}	3.552000×10^{-3}	2.336000×10^{-2}	7.008000×10^{-3}
13	2.060000×10^{-3}	6.180000×10^{-4}	1.070000×10^{-3}	3.210000×10^{-4}
17	1.240000×10^{-3}	3.720000×10^{-4}	6.400000×10^{-4}	1.920000×10^{-4}

Table A2. *Cont.*

Node	Active Power Load		Reactive Power Load	
	Expectation	Variance	Expectation	Variance
20	4.370000×10^{-3}	1.311000×10^{-3}	2.260000×10^{-3}	6.780000×10^{-4}
22	1.000000×10^{-2}	3.000000×10^{-3}	5.170000×10^{-3}	1.551000×10^{-3}
23	2.700000×10^{-2}	8.100000×10^{-3}	2.162000×10^{-2}	6.486000×10^{-3}
24	5.000000×10^{-2}	1.500000×10^{-2}	0.000000	0.000000
26	4.657000×10^{-2}	1.397100×10^{-2}	2.972000×10^{-2}	8.916000×10^{-3}
27	3.040000×10^{-3}	9.120000×10^{-4}	1.570000×10^{-3}	4.710000×10^{-4}
28	1.310000×10^{-2}	3.930000×10^{-3}	6.770000×10^{-3}	2.031000×10^{-3}
29	1.490500×10^{-1}	4.471500×10^{-2}	1.490000×10^{-2}	4.470000×10^{-3}
30	9.200000×10^{-3}	2.760000×10^{-3}	4.760000×10^{-3}	1.428000×10^{-3}
31	8.859999×10^{-3}	2.658000×10^{-3}	7.090000×10^{-3}	2.127000×10^{-3}
32	7.540000×10^{-3}	2.262000×10^{-3}	3.900000×10^{-3}	1.170000×10^{-3}
34	1.945000×10^{-2}	5.835000×10^{-3}	1.344300×10^{-1}	4.032900×10^{-2}

References

1. Guan, Z.; Tang, P.; Mao, C.; Wang, D.; Wang, L.; Liu, W.; Du, M.; Li, J.; Wang, X. Control Strategy and Implementation of Seamless Closed-Loop Load Transfer Mobile Prototype for 400 V Distribution Network. *IEEE Access.* **2024**, *12*, 12279–12294. [CrossRef]
2. Li, J.; Wang, X.; Xie, C.; Han, J.; Mao, C.; Liu, X.; Wang, D.; Zhang, Y.; Guan, Z.; Zhang, X. A novel method for seamless closed-loop load transfer in radial distribution networks. *IEEE Trans. Ind. Appl.* **2023**, *59*, 2254–2265. [CrossRef]
3. Liao, H.; Liu, D.; Huang, Y.; Zhang, Y. Load transfer capability analysis considering interconnection of distributed generation and energy storage system. *Int. Trans. Elect. Energy Syst.* **2014**, *24*, 166–177. [CrossRef]
4. Gan, G.; Wang, Z.; Zhou, Y.; Li, L.; Li, X. Reliability and economy based closed-loop operation modes for medium voltage distribution networks. *Autom. Electr. Power Syst.* **2015**, *39*, 144–150.
5. Fernandes, D.A.; Costa, F.F.; Martins, J.R.S.; Lock, A.S.; da Silva, E.R.C.; Vitorino, M.A. Sensitive load voltage compensation performed by a suitable control method. *IEEE Trans. Ind. Appl.* **2017**, *53*, 4877–4885. [CrossRef]
6. Liao, S.; Xu, J.; Sun, Y.; Gao, W.; Ma, X.; Zhou, M.; Qu, Y.; Li, X.; Gu, J.; Dong, J. Load-damping characteristic control method in an isolated power system with industrial voltage-sensitive load. *IEEE Trans. Power Syst.* **2016**, *31*, 1118–1128. [CrossRef]
7. Li, J.; Wang, Z.; Ren, J.; Xu, M.; Tian, F.; Zou, H. Feasibility analysis of loop closing operation in a distribution network based on the main network and feeder load equivalent model. *Power Syst. Prot. Control* **2023**, *51*, 107–116.
8. Shi, A.; Yang, Z.; Pang, J.; Han, S. Analysis of key factors affecting closing current of distribution network of the medium and low voltage. In Proceedings of the 2022 Annual Meeting of CSEE Study Committee of HVDC and Power Electronics (HVDC 2022), Guangzhou, China, 18-21 December 2022; pp. 374–380.
9. Zhang, K.; Feng, Y.; Yu, Y.; Guo, Q. Study of calculation schemes of currents induced by closing-loop operations in medium-voltage distribution grids. In Proceedings of the 2019 IEEE Sustainable Power and Energy Conference (iSPEC), Beijing, China, 21–23 November 2019; pp. 2488–2493.
10. Yang, B.; Wei, Y.; Zeng, Q.; Xu, J.; Zhang, F.; Huang, M. Research on system-friendly optical storage DC microgrid sending-out control strategy via MMC. In Proceedings of the 2021 11th International Conference on Power and Energy Systems (ICPES), Shanghai, China, 18–20 December 2021; pp. 501–505.
11. Abbasi, A.R.; Mohammadi, M. Probabilistic load flow in distribution networks: An updated and comprehensive review with a new classification proposal. *Electr. Power Syst. Res.* **2023**, *222*, 109497. [CrossRef]
12. Krishna, A.B.; Abhyankar, A.R. Uniform experimental design-based nonparametric quasi-monte carlo for efficient probabilistic power flow. *IEEE Trans. Power Syst.* **2023**, *38*, 2318–2332. [CrossRef]
13. Liu, M.; Xiong, Y.; Li, Q.; Murad, M.A.A.; Zhong, W. Higher-order markov chain-based probabilistic power flow calculation method considering spatio-temporal correlations. *Energies* **2025**, *18*, 1058. [CrossRef]
14. Prusty, B.R.; Jena, D. A critical review on probabilistic load flow studies in uncertainty constrained power systems with photovoltaic generation and a new approach. *Renew. Sustain. Energy Rev.* **2017**, *69*, 1286–1302. [CrossRef]
15. Hajian, M.; Rosehart, W.D.; Zareipour, H. Probabilistic power flow by monte carlo simulation with latin supercube sampling. *IEEE Trans. Power Syst.* **2012**, *28*, 1550–1559. [CrossRef]
16. Luo, G.; Chen, J.; Cai, D.; Shi, D.; Duan, X. Probabilistic assessment of available transfer capability considering spatial correlation in wind power integrated system. *IET Gener. Transm. Distrib.* **2013**, *7*, 1527–1535.

17. Gupta, N.; Daratha, N. Probabilistic three-phase load flow for unbalanced electrical systems with wind farms. *Int. J. Electr. Power Energy Syst.* **2017**, *87*, 154–165. [CrossRef]
18. Wang, X.; Gong, Y.; Jiang, C. Regional carbon emission management based on probabilistic power flow with correlated stochastic variables. *IEEE Trans. Power Syst.* **2015**, *30*, 1094–1103. [CrossRef]
19. Schellenberg, A.; Rosehart, W.; Aguado, J. Cumulant-based probabilistic optimal power flow (P-OPF) with Gaussian and gamma distributions. *IEEE Trans. Power Syst.* **2005**, *20*, 773–781. [CrossRef]
20. Liu, H.; Tang, C.; Han, J.; Li, T.; Li, J.; Zhang, K. Probabilistic load flow analysis of active distribution network adopting improved sequence operation methodology. *IET Gener. Transm. Distrib.* **2017**, *11*, 2147–2153. [CrossRef]
21. Moreira, E.; Chagas, E.; Rodrigues, A.; da Silva, M.D.G. Linear probabilistic power flow for islanded microgrids. In Proceedings of the 2020 International Conference on Probabilistic Methods Applied to Power Systems (PMAPS), Liege, Belgium, 18–21 August 2020; pp. 1–6.
22. Singh, V.; Moger, T.; Jena, D. Maximum entropy based probabilistic load flow for assessing input uncertainties and line outages in wind-integrated power systems. *Electr. Power Syst. Res.* **2025**, *244*, 111528. [CrossRef]
23. Xiao, H.; Pei, W.; Wu, L.; Li, M.; Ma, T.; Hua, W. A novel deep learning based probabilistic power flow method for Multi-Microgrids distribution system with incomplete network information. *Appl. Energy* **2023**, *335*, 120716. [CrossRef]
24. Kim, S.; Hur, J. A probabilistic modeling based on Monte Carlo simulation of wind powered EV charging stations for steady states security analysis. *Energies* **2020**, *13*, 5260. [CrossRef]
25. Huang, Y.; Rong, N.; Chen, S.; Guo, Y. Active distribution network feeder loop closing current assessment based on DK2-PL. *Electr. Power Eng. Technol.* **2024**, *43*, 133–141.

Disclaimer/Publisher's Note: The statements, opinions and data contained in all publications are solely those of the individual author(s) and contributor(s) and not of MDPI and/or the editor(s). MDPI and/or the editor(s) disclaim responsibility for any injury to people or property resulting from any ideas, methods, instructions or products referred to in the content.

energies

MDPI

Article

A Wind Power Forecasting Method Based on Lightweight Representation Learning and Multivariate Feature Mixing

Chudong Shan [1,2,*], Shuai Liu [1,2], Shuangjian Peng [1,2], Zhihong Huang [1,2], Yuanjun Zuo [1,2], Wenjing Zhang [1,2] and Jian Xiao [1,2]

1 State Grid Hunan Electric Power Company Limited Research Institute, Changsha 410000, China
2 Hunan Province Engineering Technology Research Center of Electric Power Multimodal Perception and Edge Intelligence, Changsha 410000, China
* Correspondence: whuscd@whu.edu.cn

Abstract: With the rapid development of renewable energy, wind power forecasting has become increasingly important in power system scheduling and management. However, the forecasting of wind power is subject to the complex influence of multiple variable features and their interrelationships, which poses challenges to traditional forecasting methods. As an effective feature extraction technique, representation learning can better capture complex feature relationships and improve forecasting performance. This paper proposes a two-stage forecasting framework based on lightweight representation learning and multivariate feature mixing. In the representation learning stage, the efficient spatial pyramid module is introduced to reconstruct the dilated convolution part of the original TS2Vec representation learning model to fuse multi-scale features and better improve the gridding effect caused by dilated convolution while significantly reducing the number of parameters in the representation learning model. In the feature mixing stage, TSMixer is used as the basic model to extract cross-dimensional interaction features through its multivariate linear mixing mechanism, and the SimAM lightweight attention mechanism is introduced to adaptively focus on the contribution of key time steps and optimize the allocation of forecasting weights. The experimental results conducted on actual wind farm datasets show that the model proposed in this paper significantly improves the accuracy of wind power forecasting, providing new ideas and methods for the field of wind power forecasting.

Keywords: wind power forecasting; representation learning; artificial intelligence; feature mixture; attention mechanism

1. Introduction

With the continuous growth of global energy demand and the increasing awareness of environmental protection, renewable energy has gradually become one of the focal points of international attention. Among renewable energy sources, wind energy, as a relatively mature and widely utilized form, has enormous development potential and economic benefits [1]. However, due to the complexity and randomness of wind speed, the volatility of wind power poses challenges for the operation and planning of wind power systems [2]. Therefore, accurately forecasting the time series of wind power is important. Time series forecasting of wind power can provide strong support for grid scheduling, new energy generation planning, and the operation and maintenance of wind turbines. An effective wind power forecasting model can help optimize the scheduling and regulation of wind power systems and improve system performance and reliability while also reducing energy costs and environmental pollution [3].

In wind power time series forecasting, researchers have conducted a substantial amount of research. Currently, commonly used forecasting methods include statistical time series analysis methods, Artificial Intelligence (AI) methods, and physics-based models [4]. Among these, machine learning methods such as Support Vector Machines (SVM), Artificial Neural Networks (ANN), Deep Neural Networks (DNN), and other deep learning algorithms have been widely applied in wind power forecasting [5–7]. Karijadi et al. [8] proposed a hybrid CEEMDAN-EWT deep learning method for wind power forecasting. This method uses a combination of Complete Ensemble Empirical Mode Decomposition with Adaptive Noise (CEEMDAN) and Empirical Wavelet Transform (EWT) as preprocessing techniques, where CEEMDAN is first used to decompose the original wind power data into several sub-sequences, and the EWT denoising technique is used to denoise the highest frequency sequence generated by CEEMDAN. Then, use Long-Short Term Memory (LSTM) to predict all subsequences in the CEEMDAN-EWT process, and aggregate the prediction results of each subsequence to achieve the final prediction result. However, this method did not effectively utilize the various multivariate features that affect wind power generation. Chen et al. [9] proposed a CNN-BiLSTM short-term wind power forecasting method based on feature selection. Although this method utilizes multivariate features, it requires a feature correlation analysis of the dataset and weighting the input data based on feature correlation to form a multidimensional feature dataset, which is a cumbersome process.

Currently, traditional wind power forecasting models face numerous issues when handling time series data, such as weak feature extraction capabilities and significant information loss. In existing wind power forecasting methods, the use of feature representation is relatively limited. There are currently many methods for feature selection, such as ElasticNet, Pearson's correlation, and ReliefF algorithms [10,11], but using deep learning techniques for automatic feature extraction and representation would be more convenient and accurate. Feature representation can extract latent patterns and characteristics from time series data, thereby improving the performance of forecasting models. Traditional time series forecasting methods often rely on simple statistical features or directly use raw data, failing to fully mine the deep-seated information within the data. This leads to certain limitations for the models when confronted with the complex variations in wind power.

To address these issues, some researchers have proposed unsupervised representation learning models in recent years, among which the TS2Vec model is a typical example [12]. The TS2Vec model extracts features by mapping time series data to a low-dimensional vector space, utilizing traditional dilated convolution networks for feature processing. Although the TS2Vec model has achieved certain results in various downstream machine learning tasks, its dilated convolution module has limitations in capturing long-term dependencies, and its gridding effect may result in information loss or ambiguity. Additionally, the model makes a large number of parameters and computational complexity. The efficient spatial pyramid structure proposed in ESPNet [13] can better compensate for the gridding effect caused by dilated convolution while reducing model complexity.

It is equally a challenging problem to select suitable downstream forecasting models for multi feature mixing and extraction of data with multiple feature variables after feature representation by representation learning models. Many current time series forecasting models are based on univariate time forecasting, while multivariate models seem to have overfitting issues, especially when the target time series is not correlated with other covariates [14]. However, when there is a strong correlation between the target time series of wind power and various other meteorological factors covariates, it is meaningful to have a model that can use information from covariates and other features for forecasting [15]. In recent years, scholars have proposed different models such as PatchTST [16], DLinear [17], CycleNet [18], etc., which are based on channel independent implementation. These models usually do not consider any potential interactions or correlations between channels, and each channel is processed as a separate input without utilizing shared information or dependencies [19]. Various time series foundation models, such as LLM4TS [20], Time LMM [21], Chronos [22], etc., are also based on channel independence. Although LLM4TS and Time LLM introduce pre-training techniques, their self attention mechanisms are still limited to single variable channels, making it difficult to establish dynamic weight allocation mechanisms across feature domains.

In response to the above bottlenecks, multivariate feature mixing has become a new direction to break through the limitations of channel independence [23]. TSMixer, as a novel architecture, effectively captures temporal patterns and cross-variable information through temporal and feature mixing operations, enabling dynamic fusion of multivariate time series [24]. However, existing research still faces challenges in designing downstream forecasting tasks based on feature representation, where redundancy and noise coexist in high-dimensional feature spaces, and the contribution of key covariates needs to be strengthened through adaptive weight allocation [25]. SimAM, as a lightweight and parameter-free attention mechanism, generates attention weights by calculating the local self-similarity of feature maps without introducing any additional parameters [26]. The introduction of a lightweight SimAM attention mechanism can evaluate feature importance through an energy function that does not require parameter optimization, and can highlight the weights of key segments in multivariate time series under controllable computational costs [27].

Therefore, this paper proposes a wind power forecasting method and constructs a two-stage forecasting framework based on lightweight representation learning and multivariate feature mixing. In the representation learning stage, the efficient spatial pyramid module is utilized to replace the dilated convolution module in the original TS2Vec model. The improved TS2Vec representation model is utilized to characterize the matrix containing historical measured weather data, historical power generation data, and other features of the wind farm. The multivariate feature data after representational learning serves as input for the downstream forecasting model. In the stage of multivariate feature mixing, the use of SimAM-TSMixer hybrid architecture is intended to break through the inherent limitations of channel-independent models and provide a new solution for multivariate wind power forecasting. By efficiently combining SimAM with TSMixer, the model can dynamically calibrate the correlation strength between meteorological covariates and target time series during feature mixing, while mining spatiotemporal interaction patterns across variables.

The main contributions of this paper are as follows:

1. A two-stage forecasting framework based on lightweight representation learning and multivariate feature mixing is proposed, which can effectively extract potential patterns and features in wind power related time series, and at the same time can efficiently realize the dynamic fusion of multivariate features, enabling the model to better adapt to complex time series data.
2. In the lightweight representation learning stage, the dilated convolution part of the existing TS2Vec representation model is innovatively modified, and the efficient spatial pyramid structure is adopted, which better compensates for the gridding effect caused by the dilated convolution. This not only enhances the ability of the model to capture multi-scale features, but also improves the flexibility and adaptability of the model in dealing with complex time series data.
3. In the multivariate feature mixing stage, a multivariate mixing layer is constructed based on the TSMixer architecture, which utilizes its cross-dimensional interaction

mechanism to extract implicit associations among features, and embeds the SimAM lightweight attention mechanism, which adaptively adjusts the weights of the time steps through parameter-free computation to suppress the noise interference and enhance the contribution of key features.

2. Data

To verify the effectiveness and practicality of the model proposed in this paper, the data used in the study consists of real historical data from a wind farm in Hunan Province, China, over the course of one year. The data sampling frequency is 15 min and includes measurements such as wind speed and direction at 10 m, wind speed and direction at 50 m, hub wind speed and direction, temperature, pressure, and historical actual power.

Due to the long-term operation of wind power generation systems and their supporting meteorological monitoring equipment, various problems may occur in data collection (such as equipment failures, signal interference, network transmission interruptions, etc.), resulting in noise, erroneous records, and missing data in the original data. The main purpose of data cleaning is to eliminate invalid data, fill in missing values, and correct erroneous records, thereby ensuring the integrity and accuracy of the dataset. Due to the uncertainty of missing values, the longest continuous missing time is approximately 3 h, which means 12 consecutive sampling points. The Isolation Forest algorithm [28] is used for outlier detection, to fill in missing values and solve outliers, we used the commonly used linear interpolation [29] method as shown in Equation (1), where ($y_2 > y_1$) and ($x_2 > x > x_1$). In the equation, y represents the value of the unknown data point to be estimated, and x is the value of the unknown data point. This equation is based on the principle of similar triangles, using the two known data points ($x_1,\ y_1$) and ($x_2,\ y_2$) to calculate the data of the interpolated point.

$$y = y_1 + \frac{(y_2 - y_1) \times (x - x_1)}{x_2 - x_1} \tag{1}$$

We use StandardScaler for data normalization [30], and the formula for data normalization is shown in Equation (2):

$$Z = \frac{x\ -\ \mu}{\sigma} \tag{2}$$

Here x is the original data value, μ is the mean of the feature, which is the average of all data points, and σ is the standard deviation of the feature, which indicates the dispersion of the data points.

3. Methods and Results

3.1. Lightweight Representation Learning Model

3.1.1. Model Design

The lightweight representation learning model set out in the present paper enhances the dilated convolution in the encoder part of the original TS2Vec model. As illustrated in Figure 1, the original TS2Vec model employs a dilated convolution module with 11 residual blocks to extract contextual representations for each timestamp. Each residual block contains two one-dimensional dilated convolution layers with a dilation parameter of 2^{i-1} (i represents the order of the residual blocks) and a kernel size of 3. The dilation parameter of the dilated convolution in each residual block increases from 1 to 1024. However, each residual block contains two dilated convolutions with the same dilation parameter, which can lead to some pixels in the feature map not participating in the computation when dilated convolutions with the uniform dilation parameter are stacked. Additionally, the gridding effect of the dilated convolution can result in a lack of interdependence between the results of convolution layers, leading to insufficient correlation between the convolution results of corresponding layers and causing local information loss. Although the original TS2Vec model uses a residual structure to overcome some gridding effects caused by dilated convolutions, the dilated convolution residual block in the model performs convolution operations based on the aforementioned residual block rather than separately extracting multi-scale features from the original input data. Furthermore, the large number of convolution layers in the model results in a significant number of parameters and computational load. This paper refers to the scheme proposed in ESPNet [13] to replace the dilated convolution module in the original TS2Vec model with the efficient spatial pyramid module in Figure 2.

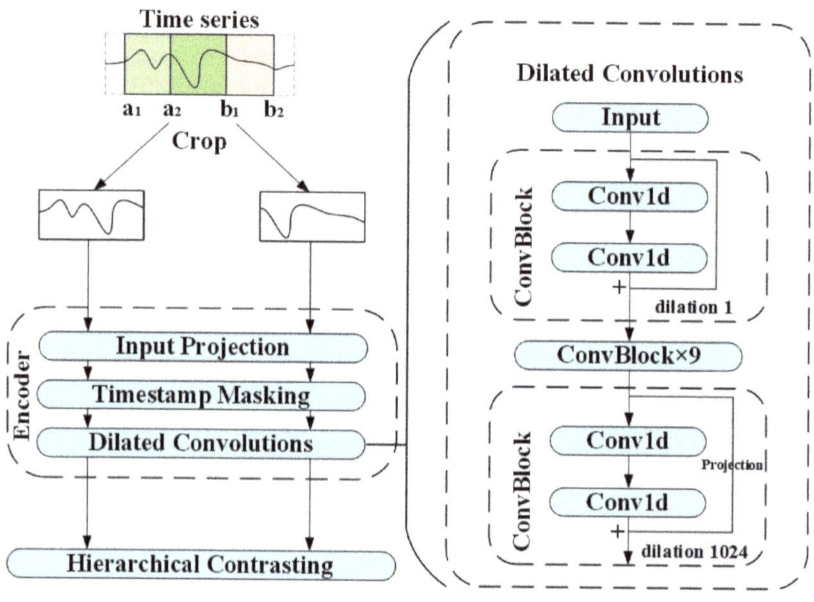

Figure 1. Original TS2Vec model schemes.

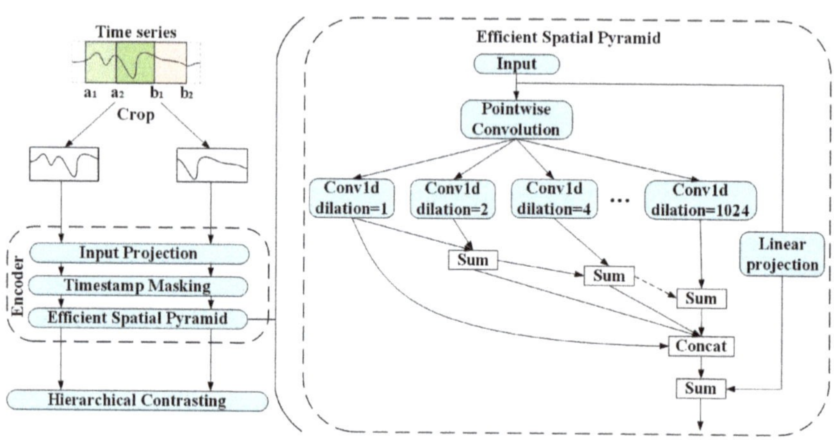

Figure 2. Structure of the lightweight representation learning model.

Figure 2 illustrates the structure of the lightweight representation learning model set out in the present paper. The starboard side of Figure 2 shows the specific implementation method of the efficient spatial pyramid module. To reduce the channel number of the output feature map after the timestamp masking module in the original TS2Vec model and thereby decrease the subsequent computational load, a one-dimensional pointwise convolution with a kernel size of 1 is applied. Additionally, 11 parallel dilated convolution layers with dilation parameters ranging from 1 to 1024 are utilized to perform convolution operations on the feature map after pointwise convolution. Each dilated convolution layer has a dilation parameter of 2^{i-1} (where i represents the layer number), and the padding value is consistent with the dilation parameter of that layer. The feature map output from the dilated convolution layer with the smallest dilation parameter is progressively stacked with the output feature maps from each dilated convolution layer, and each stacked feature map is concatenated. Finally, the concatenated feature map is added to the initially linearly projected input feature map to form a residual structure. This structure employs the concept of feature hierarchy to directly utilize dilated convolutions with different dilation parameters for multi-scale feature extraction on the input feature map after pointwise convolution, layering, and stacking the output of dilated convolutions with different dilation parameters. Essentially, this adds discrete receptive fields, effectively compensating for the gridding effect caused by dilated convolutions, preserving local details and global semantic features, and better capturing multi-scale temporal information. Meanwhile, the use of pointwise convolution and a reduced number of convolution layers decreases the overall computational load of the model.

The hierarchical contrasting module in Figures 1 and 2 utilizes temporal contrastive loss and instance-wise contrastive loss values to capture the contextual representation of the time series. The representation of the same timestamp from two views of the input time series is considered positive, while the representation of different timestamps from the same time series is considered negative. Let i be the index of the input time series sample and t be the timestamp. The temporal contrastive loss of the i-th time series with timestamp t can be given in Equation (3).

$$\ell_{temp}^{(i,t)} = -\log \frac{\exp\left(r_{i,t} \cdot r'_{i,t}\right)}{\sum_{t' \in \Omega}\left(\exp\left(r_{i,t} \cdot r'_{i,t'}\right) + 1_{[t \neq t']}\exp\left(r_{i,t} \cdot r_{i,t'}\right)\right)} \tag{3}$$

The instance-wise contrastive loss indexed by (i, t) can be expressed as Equation (4).

$$\ell_{inst}^{(i,t)} = -\log \frac{\exp\left(r_{i,t} \cdot r'_{i,t}\right)}{\sum_{j=1}^{B}\left(\exp\left(r_{i,t} \cdot r'_{j,t}\right) + \mathbb{1}_{[i \neq j]}\exp\left(r_{i,t} \cdot r_{j,t}\right)\right)} \tag{4}$$

3.1.2. Model Testing and Result Analysis

To validate the performance of the lightweight representation learning model set out in the present paper, we use the feature matrix of wind farm historical data after feature representation combined with the downstream time series forecasting model LSTM [31] to test the lightweight representation learning model. Figure 3 displays the specific construction method of the LSTM model used in the model testing process. The LSTM model consists of 5 layers, with the first 3 layers of LSTM hidden layers having a size of units set to 50 and a return sequence status set to True. The fourth layer of LSTM hidden layers has a size of units set to 50 and a return sequence status set to False. The fifth layer is the fully connected Dense layer, and the output data dimension is placed at 96, corresponding to a data length of one day in the dataset.

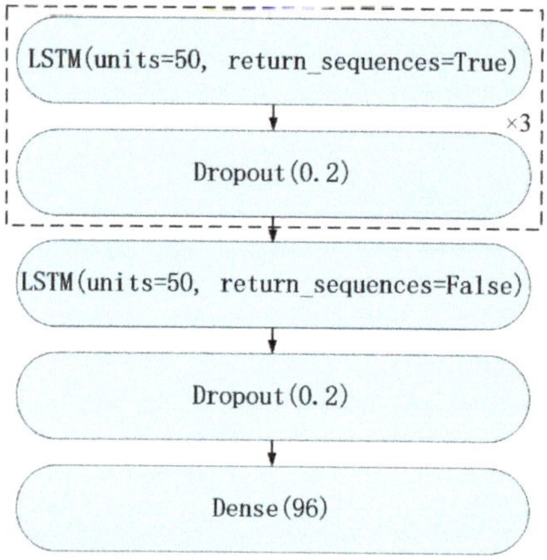

Figure 3. Details of LSTM during model testing process.

We first use the LSTM model to make downstream forecasting on the actual data of the wind farm represented by the original TS2Vec model and the actual data of the wind farm represented by the improved TS2Vec model, which is our proposed lightweight representation learning model. The dataset includes wind speed at 10 m, wind direction at 10 m, wind speed at 50 m, wind direction at 50 m, hub height wind speed, hub height wind direction, temperature, air pressure, and historical actual power data from wind turbines within one year. The dataset size is (35,040, 9).

As shown in Figure 1, the original TS2Vec model's dilated convolution part is set to have a kernel size of 3, and the first 10 residual blocks have 64 input and output feature map channels, while the last residual block has 352 output feature map channels. For the improved TS2Vec model, since the original TS2Vec model outputs a feature map with a feature dimension of 64 after the timestamp masking module, the feature map is transposed to the number of channels of 64. Then, a one-dimensional pointwise convolution with a kernel size of 1 is utilized to reduce the number of channels in the feature map to 32, thereby reducing the subsequent computational load. In the subsequent parallel dilated convolution section, the kernel size of each parallel dilated convolution layer is set as 3, and the number of channels in the output feature map is 32. The feature maps output by each dilated convolution layer are stacked step by step, and each stacked feature map is concatenated. Finally, the concatenated feature map is added to the input feature map with a channel number of 352 that was first projected through a linear connection layer to obtain the output feature map with a channel number of 352. Using 70% of the data in the dataset for training the original TS2Vec model and the improved TS2Vec model, and encoding the original dataset with the trained models, a representation dataset of size (35,040, 352) can be achieved.

As for the LSTM model for downstream time series forecasting used for model testing, a historical data set with a length of 288, which corresponds to three days of data after representation learning, is set as input x. The actual power of the original data with a length of 96, which corresponds to one day, is taken as output y to train the LSTM model. Similarly, 70% of the original data set is used as the training set, 10% of the data set is used as the validation set, and 20% of the data set is used as the testing set.

The evaluation indicators of the original TS2Vec model and the improved TS2Vec model combined with LSTM for time series forecasting on the test set are as follows [32], MAE and RMSE have the same unit as y.

$$\mathrm{MAE} = \frac{1}{m}\sum_{i=1}^{m} | (y_i - \hat{y}_i) | \tag{5}$$

$$\mathrm{MSE} = \frac{1}{m}\sum_{i=1}^{m} \left(y_i - \hat{y}_i\right)^2 \tag{6}$$

$$\mathrm{RMSE} = \sqrt{\frac{1}{m}\sum_{i=1}^{m} (y_i - \hat{y}_i)^2} \tag{7}$$

The training and testing experiments of the model were conducted on an x86-64 Linux operating system equipped with Intel (R) xeon (R) Gold 6430 CPU and NVIDIA A800 SXM4 80 GB GPU, using the PyTorch 2.1.0 and CUDA 12.1 environment.

Table 1 displays the test results of the lightweight representation learning model proposed in this paper, namely the improved TS2Vec model and the original TS2Vec model, when combined with the same downstream forecasting model LSTM for time series forecasting on the same dataset.

Table 1. Comparison of performance testing between the original TS2Vec model and the improved TS2Vec model.

Model	MAE	MSE	RMSE
Original TS2Vec + LSTM	0.6438	0.7305	0.8547
Improved TS2Vec + LSTM	**0.6228**	**0.7268**	**0.8525**

We simultaneously calculated the parameter and computational complexity of the Encoder section of both the original TS2Vec model and the improved TS2Vec model on the test set [33]. The computational complexity (FLOPs) stands for floating point operations per second, the parameter count (Params) is mainly used to designate the size of the model. The comparison results are presented in Table 2.

Table 2. Comparison of model size between the original TS2Vec model and the improved TS2Vec model.

Model	FLOPs	Params
Original TS2Vec + LSTM	4.9659 G	0.7110 M
Improved TS2Vec + LSTM	**0.4286 G**	**0.0609 M**

From Table 2, it can be observed that compared to the original TS2Vec model, the improved TS2Vec model combined with LSTM significantly reduces computational and parameter complexity while ensuring forecasting accuracy.

We also compared the results of time series forecasting on the original dataset using the LSTM model and several other existing deep learning time series models without using representation learning models, with the forecasting results of the improved TS2Vec set out in the present paper combined with the LSTM model. The results are presented in Table 3. Meanwhile, Figures 4–7 demonstrate the performance of the improved TS2Vec combined with the LSTM model and other mainstream deep learning time series forecasting models on the same test set.

From the comparison curves of the improved TS2Vec combined with LSTM model and other mainstream deep learning time series forecasting models' wind power forecasting results and actual wind power generation values in randomly selected periods shown in Figures 4–7, it can be observed that the improved TS2Vec combined with LSTM model's forecasting results are closest to the true curve. From the above results, it be observed that the use of the representation learning model has significantly improved the wind power time series forecasting results. At the same time, the improved representation model has significantly reduced the parameter and computational complexity compared to the original TS2Vec model, which well proves the superiority of the lightweight representation learning model proposed in this paper.

Table 3. Comparison of performance testing between the improved TS2Vec combined with LSTM model and other mainstream deep learning models.

Model	MAE	MSE	RMSE
MLP [34]	0.7178	0.9589	0.9793
RNN [35]	0.7538	0.8680	0.9317
N-BEATS [36]	0.7001	0.8787	0.9374
LSTM	0.7380	0.9790	0.9895
Original TS2Vec + LSTM	0.6438	0.7305	0.8547
Improved TS2Vec + LSTM	**0.6228**	**0.7268**	**0.8525**

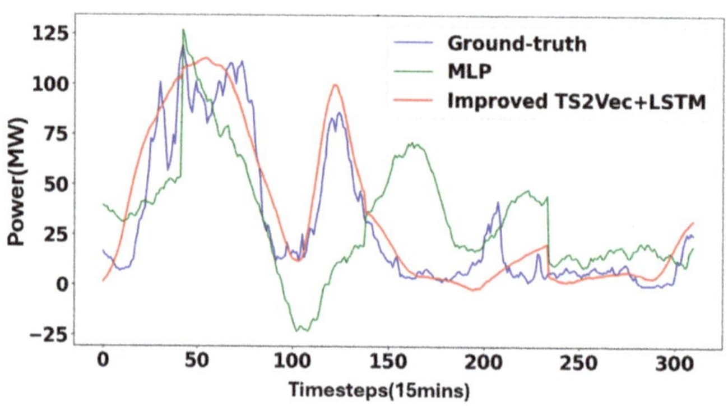

Figure 4. Comparison between improved TS2Vec combined with LSTM and MLP.

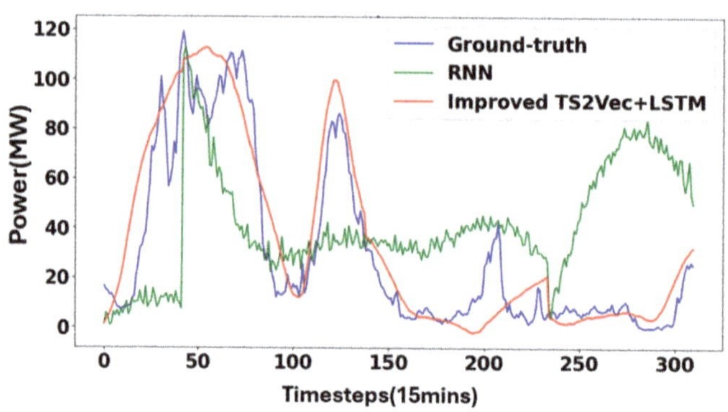

Figure 5. Comparison between improved TS2Vec combined with LSTM and RNN.

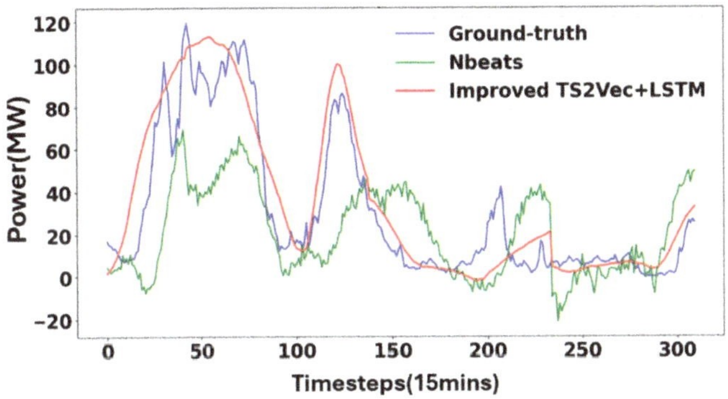

Figure 6. Comparison between improved TS2Vec combined with LSTM and N-BEATS.

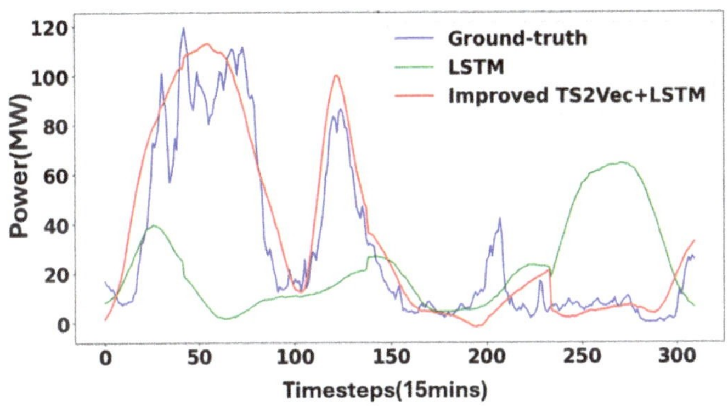

Figure 7. Comparison between improved TS2Vec combined with LSTM and original LSTM.

3.2. Multivariate Feature Mixing Model

3.2.1. Model Design

In the earlier part of the paper, we designed a lightweight representation learning model and verified through experiments that the feature representation model achieved good representation performance. As the original data dimension after feature representation became 352 dimensions, we studied how to develop a better multivariate feature mixing downstream time series forecasting model that can fully integrate and extract the features of each dimension in multivariate time series data. As entered in Section 1, many current time series forecasting models use the idea of channel independence. Therefore, we draw inspiration from the Time Series Mixer (TSMixer) [24] model and use a stacked multilayer perceptron architecture to achieve cross-variable information mixing and feature extraction of high-dimensional and multivariate historical weather and power data of wind farms after feature representation, thereby further improving the accuracy of wind power forecasting. In addition, we have introduced a lightweight, parameter-free attention mechanism, SimAM [25], based on downstream forecasting models, which generates attention weights by calculating the local self-similarity of feature maps without introducing any additional parameters, thus improving the performance of downstream forecasting models.

The overall structure diagram of the multivariate feature mixing downstream forecasting model designed in this paper is presented in Figure 8. In terms of the dataset, we use the improved TS2Vec representation model to represent the data with a size of (35,040, 352) as the dataset. At the same time, to further highlight the impact of historical power generation on future power generation, we also concatenated historical power generation data in the last column based on 352-dimensional data, so the size of the dataset becomes (35,040, 353). The model uses historical data from the past three days to forecast the power generation for the next day. Here, a length of 288 is configured as the input, which corresponds to the size of (288, 353) after representation learning. The output size of the model is (96, 1), which is the power generation for the next day.

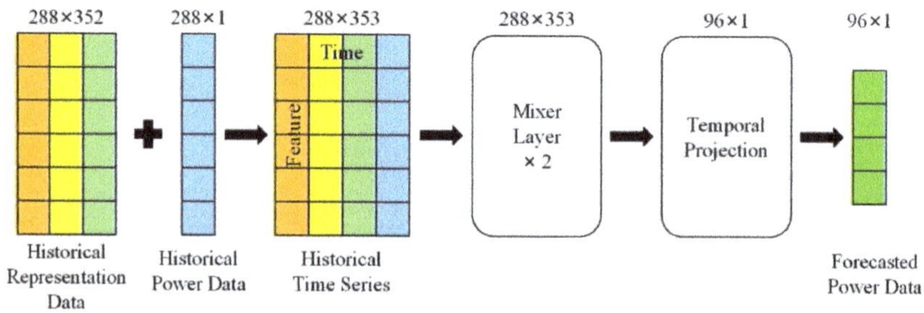

Figure 8. The overall structure of multivariate feature mixing model.

From Figure 8, it can be observed that the multivariate feature mixing model includes two stacked Mixer Layer modules, which are the core modules of the multivariate feature mixing model. Figure 9 shows the specific implementation details of the mixer layer in the multivariate feature mixing model. The mixer layer is mainly composed of time mixing and feature mixing, which can fuse the time and feature levels of input data to better forecast time series using more information from the input data. Both the time mixing and feature mixing modules include the MLP module. For time mixing, MLP_{time} consists of a fully connected layer, a ReLU activation function, and a dropout layer. The input feature map (where rows represent time and columns represent features) is first transposed, and MLP_{time} is implemented in the time domain and shared among all features. For feature mixing, $MLP_{feature}$ consists of two fully connected layers, ReLU activation function, and dropout layer. $MLP_{feature}$ applies to the feature domain and is shared across all time steps. The time mixing and feature mixing modules can automatically adapt to the use of time and cross-variable information, and both modules utilize residual connections to help the model learn deeper data representations while maintaining reasonable computational costs [37].

The attention mechanism can automatically identify important parts of the input data, highlight key features through weight allocation, and reduce the interference of redundant information on the subsequent processing of the MLP. This filtering mechanism significantly improves model efficiency, especially in complex inputs such as long sequences and multimodal data [38]. From the Figure 9, to enable the model to better extract the features of the original data in the time and feature dimensions, this paper introduces the lightweight and parameter free attention mechanism module SimAM, which has good application effects in the image field, in both time mixing and feature mixing, and modifies its structure to an attention mechanism module suitable for one-dimensional time series domain.

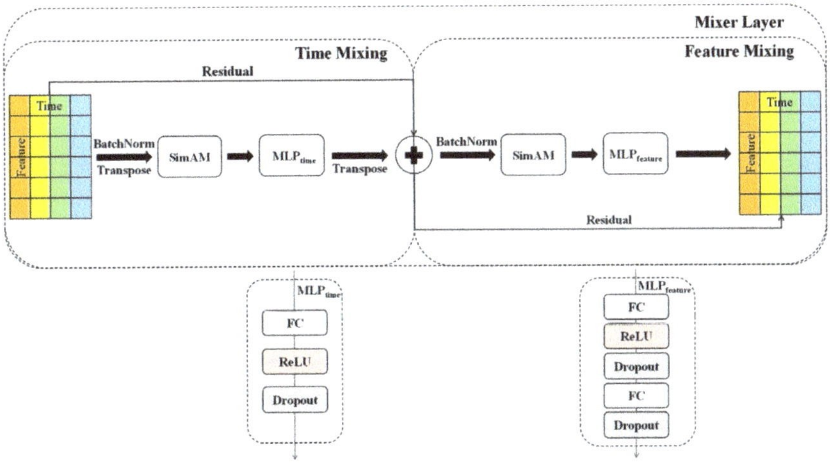

Figure 9. The specific implementation details of the mixer layer.

As indicated in Figure 10, many attention mechanism modules currently operate along the channel or spatial dimension, generating channel or spatial weights and treating neurons in each channel or spatial position equally [39]. SimAM can generate both channel and spatial weights simultaneously, generating attention weights by calculating the similarity between each pixel in the feature map and its adjacent pixels. For the input feature map $X \in \mathbb{R}^{B\times C\times L}$, where B represents batch size, C represents the number of channels in the feature map, and L represents the length of the one-dimensional time series feature map. For each pixel x_i in the feature map, as shown in Equation (8), SimAM first indirectly reflects similarity by calculating the average square of the difference between x_i and its neighboring pixels. Ω_i represents the neighborhood of x_i, and N represents the number of pixels in the neighborhood. However, to simplify the calculation, this paper uses the average of pixels in the feature map as x_k to calculate s_i.

$$s_i = \frac{1}{N}\sum_{k\in\Omega_i} |x_i - x_k|^{2} \quad (8)$$

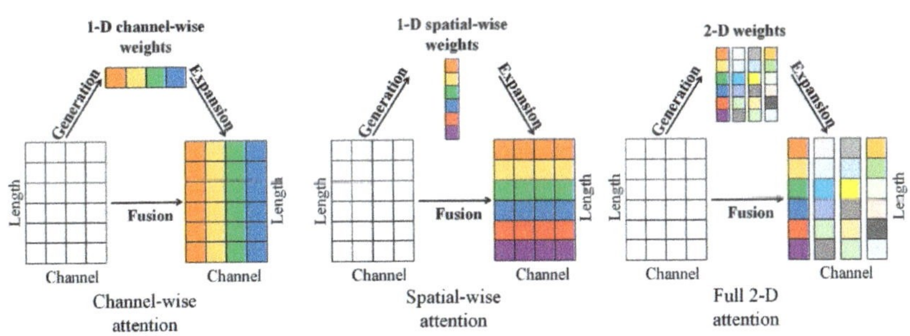

Figure 10. Comparison of different types of attention mechanisms for time series.

After obtaining s_i, the attention weight w_i is calculated using Equation (9), which is similar to the sigmoid function that projects the weight values into the (0, 1) interval, where σ_i^2 is the normalization of s_i and ϵ is taken as 1×10^{-4}. After obtaining the attention weight map, it can be multiplied element by element with the original feature map to obtain the feature map after passing through the attention module.

$$w_i = \frac{1}{1 + \exp\left(-\frac{1}{4}\left(\frac{s_i}{\sigma_i^2+\epsilon} - 1\right)\right)} \quad (9)$$

Figure 11 displays the temporal projection module in the multivariate feature mixing model of this paper. The module first takes the data from the target column of the feature map processed by the mixer layer. After various time mixing, feature mixing, and SimAM attention mechanism modules in the early stage, the target column not only retains the information of historical power generation, but also integrates numerous cross features from other dimensions and spaces. After being applied to the fully connected layer in the time domain, it can learn the time pattern and map the time series from the original input length of 288 to the target forecasting length of 96.

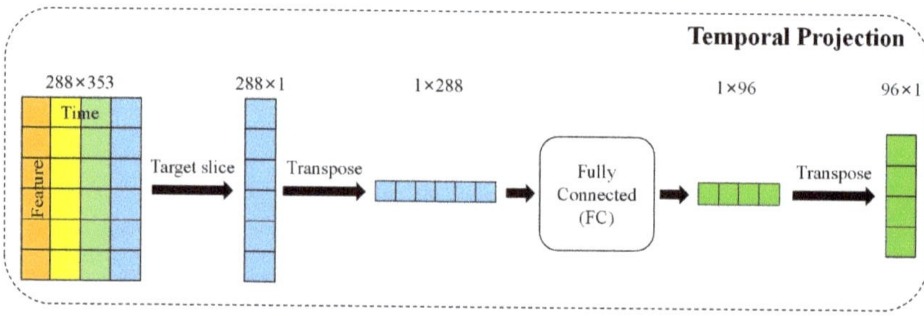

Figure 11. The specific implementation details of the temporal projection module.

3.2.2. Model Testing and Result Analysis

In order to verify the compatibility between the multivariate feature mixing model and the lightweight representation learning model proposed in this paper, as well as the forecasting performance of the model, we conducted sufficient experiments. The training and testing experiments of the model were conducted on an x86-64 Linux operating system equipped with Intel (R) xeon (R) Gold 6430 CPU and NVIDIA A800 SXM4 80 GB GPU, using the Pytorch framework.

We took the lightweight representation learning model for feature representation and used 70% of the data in the 35,040 length dataset as the training set, 10% of the data as the validation set, and 20% of the data as the testing set to train and test the performance of the multivariate feature mixing model and compare them with other mainstream models. At the same time, ablation experiments were designed to verify the contribution of the SimAM attention module in a multivariate feature mixing model to the improvement of model forecasting performance. Utilizing MAE, MSE, and RMSE as evaluation indicators, the Table 4 shows the specific performance of the model on different indicators.

From Table 4, it can be observed that we conducted experiments on various combinations and found that adding the SimAM attention module to the downstream forecasting model resulted in a decrease in forecasting error. In particular, the comparison data in the third and fourth rows of Table 4 shows that the SimAM module has a particularly significant improvement in forecasting performance when the representation model is not used on the original data. Even with the use of the SimAM module for the data after feature representation, the forecasting performance of the model can be enhanced to a certain extent. In addition, we compared the experimental results of the original TS2Vec model combined with our proposed multivariate feature mixing downstream forecasting model and the improved TS2Vec model combined with our proposed downstream forecasting model in this paper. It was found that the improved TS2Vec model combined with the downstream forecasting model performed better, further confirming the effectiveness of the multivariate feature mixing model in this paper. At the same time, we also listed the experimental results of the improved TS2Vec model combined with SimAM attention module and the currently advanced lightweight time series forecasting model SparseTSF [40] in Table 4. Through comparison, we can further verify the advantages of the model scheme used throughout this paper. At the same time, Table 3 also lists the performance of other models on the same dataset, all of which have well-verified the progressive nature of the model set out in the present paper. Figure 12 shows the comparison results between the forecasted power and actual power of the wind power forecasting model proposed in this paper on a long-term test set. It can be observed that the trend of the model's forecasted results and true values is similar in most intervals. The model proposed in this paper not only has significant advantages in the evaluation indicators such as MAE, MSE, RMSE listed in Table 4, but also demonstrates the effectiveness and accuracy of the proposed model in wind power forecasting by maintaining a consistent trend between the visualized wind power forecasting results and the actual values over a long period of time, as shown in Figure 12.

Table 4. Comparison of wind power forecasting performance on various combinations.

Model	MAE	MSE	RMSE
Improved TS2Vec + LSTM	0.6228	0.7268	0.8525
Improved TS2Vec + SimAM + SparseTSF [40]	0.6674	0.7202	0.8486
TSMixer	0.7053	0.8167	0.9037
SimAM + TSmixer	0.6673	0.7423	0.8615
Original TS2Vec + TSMixer	0.5404	0.4889	0.6992
Original TS2Vec + SimAM + TSMixer	0.5334	0.4721	0.6871
Improved TS2Vec + TSMixer	0.3780	0.2477	0.4977
Improved TS2Vec + SimAM + TSmixer	**0.3735**	**0.2434**	**0.4934**

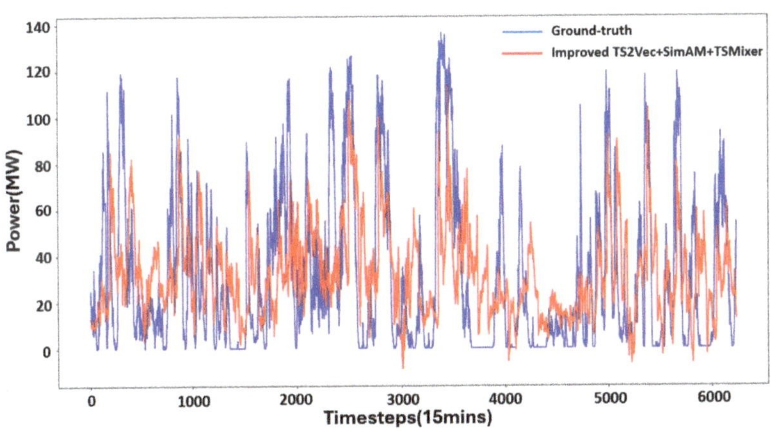

Figure 12. Comparison between the forecasted results and actual wind power of our proposed model on a long-term test set.

4. Discussion and Future Work

This paper proposes an innovative two-stage forecasting framework for wind power forecasting tasks, which achieves improved forecasting performance through a collaborative optimization mechanism of lightweight representation learning and multivariate feature mixing. In the representation learning stage, by designing the efficient spatial pyramid module to reconstruct the multi-scale feature fusion path, the gridding effect caused by traditional dilated convolution is effectively alleviated, and the model parameters and computational complexity are compressed, significantly improving the efficiency of feature extraction. In the feature mixing stage, the TSMixer model achieves deep mining of cross-dimensional interactive features through a linear mixing layer, and combines the SimAM attention mechanism to construct a dynamic weight allocation network, enabling the model to adaptively capture the contribution differences of key time steps. The experimental results show the rationality and effectiveness of the model design. Future research will focus on the fusion modeling of multimodal meteorological data, exploring a hybrid modeling paradigm of physical constraints and data-driven approaches to enhance predictive robustness under extreme weather conditions, and attempting to extend the framework to a wider range of time series forecasting scenarios such as photovoltaic power forecasting and power load forecasting, promoting the development of intelligent scheduling technology for clean energy.

5. Conclusions

This study presents a two-stage wind power forecasting framework integrating lightweight representation learning and multivariate feature mixing, which significantly improves forecasting accuracy through the collaborative optimization of an improved TS2Vec model and a SimAM-TSMixer hybrid architecture. Experimental results show that the proposed model outperforms existing methods across key metrics (MAE, MSE, and RMSE), validating its effectiveness. The efficient spatial pyramid module in the representation learning stage mitigates the gridding effect of dilated convolution while reducing computational costs, whereas the feature mixing stage leverages TSMixer's cross-dimensional interaction and SimAM's parameter-free attention to dynamically weight critical time steps. However, limitations remain: (1) The model's adaptability to high-frequency fluctuating data requires further validation. (2) The linear assumptions in feature mixing may constrain nonlinear modeling under extreme weather conditions. In realistic applications, ensuring data quality and timeliness is crucial, and site-specific parameter tuning is recommended to optimize performance. Additionally, the framework could be expanded to broader time-series forecasting scenarios, such as photovoltaic power and load forecasting, to advance intelligent scheduling technologies for clean energy.

Author Contributions: Conceptualization, C.S.; methodology, C.S.; software, C.S.; validation, C.S.; formal analysis, S.L. and Z.H.; investigation, S.L. and Z.H.; resources, S.L. and Y.Z.; data curation, S.P. and Y.Z.; writing—original draft preparation, S.P.; writing—review and editing, S.P.; visualization, W.Z.; supervision, W.Z.; project administration, J.X.; funding acquisition, J.X. All authors have read and agreed to the published version of the manuscript.

Funding: This research was funded by Science and Technology Project of State Grid Corporation of China Headquarters, titled "Key Technologies and Applications of Multi temporal and Spatial Scale Prediction of Power Source Load for Public Services Based on Pretrained Models", grant number 5216A5240018.

Data Availability Statement: The datasets presented in this article are not readily available due to commercial restrictions and data privacy protection.

Conflicts of Interest: Authors Chudong Shan, Shuai Liu, Shuangjian Peng, Zhihong Huang, Yuanjun Zuo, Wenjing Zhang and Jian Xiao were employed by State Grid Hunan Electric Power Company Limited Research Institute.

Abbreviations

The following abbreviations are used in this manuscript:

AI	Artificial Intelligence
SVM	Support Vector Machine
ANN	Artificial Neural Network
DNN	Deep Neural Network
CEEMDAN	Complete Ensemble Empirical Mode Decomposition with Adaptive Noise
EWT	Empirical Wavelet Transform
LSTM	Long-Short Term Memory
ESPNet	Efficient Spatial Pyramid Net
PatchTST	Patch Time Series Transformer
SimAM	Similarity-Aware Activation Module
LLM4TS	Large Language Models for Time Series
LLM	Large Language Model
TSMixer	Time Series Mixer
MLP	Multilayer Perceptron
RNN	Recurrent Neural Network
FLOPs	Floating Point Operations
CPU	Central Processing Unit
GPU	Graphics Processing Unit
MAE	Mean Absolute Error
MSE	Mean Squared Error
RMSE	Root Mean Square Error

References

1. Nazir, M.S.; Wang, Y.; Bilal, M.; Abdalla, A.N. Wind energy, its application, challenges, and potential environmental impact. In *Handbook of Climate Change Mitigation and Adaptation*; Springer: Cham, Switzerland, 2022; pp. 899–935.
2. Ahmed, S.D.; Al-Ismail, F.S.; Shafiullah, M.; Al-Sulaiman, F.A.; El-Amin, I.M. Grid integration challenges of wind energy: A review. *IEEE Access* **2020**, *8*, 10857–10878. [CrossRef]
3. Tsai, W.-C.; Hong, C.-M.; Tu, C.-S.; Lin, W.-M.; Chen, C.-H. A review of modern wind power generation forecasting technologies. *Sustainability* **2023**, *15*, 10757. [CrossRef]
4. Qiao, Y.; Lu, Z.; Min, Y. Research & application of raising wind power prediction accuracy. *Power Syst. Technol.* **2017**, *41*, 3261–3268.
5. Zheng, Y.; Ge, Y.; Muhsen, S.; Wang, S.; Elkamchouchi, D.H.; Ali, E.; Ali, H.E. New ridge regression, artificial neural networks and support vector machine for wind speed prediction. *Adv. Eng. Softw.* **2023**, *179*, 103426. [CrossRef]
6. Ateş, K.T. Estimation of short-term power of wind turbines using artificial neural network (ANN) and swarm intelligence. *Sustainability* **2023**, *15*, 13572. [CrossRef]
7. Tarek, Z.; Shams, M.Y.; Elshewey, A.M.; El-kenawy, E.-S.M.; Ibrahim, A.; Abdelhamid, A.A.; El-dosuky, M.A. Wind Power Prediction Based on Machine Learning and Deep Learning Models. *Comput. Mater. Contin.* **2022**, *74*, 715–732. [CrossRef]
8. Karijadi, I.; Chou, S.-Y.; Dewabharata, A. Wind power forecasting based on hybrid CEEMDAN-EWT deep learning method. *Renew. Energy* **2023**, *218*, 119357. [CrossRef]
9. Chen, Y.; Zhao, H.; Zhou, R.; Xu, P.; Zhang, K.; Dai, Y.; Zhang, H.; Zhang, J.; Gao, T. CNN-BiLSTM short-term wind power forecasting method based on feature selection. *IEEE J. Radio Freq. Identif.* **2022**, *6*, 922–927. [CrossRef]
10. Yousaf, S.; Bradshaw, C.R.; Kamalapurkar, R.; San, O. Investigating critical model input features for unitary air conditioning equipment. *Energy Build.* **2023**, *284*, 112823. [CrossRef]
11. Yousaf, S.; Bradshaw, C.R.; Kamalapurkar, R.; San, O. A gray-box model for unitary air conditioners developed with symbolic regression. *Int. J. Refrig.* **2024**, *168*, 696–707. [CrossRef]
12. Yue, Z.; Wang, Y.; Duan, J.; Yang, T.; Huang, C.; Tong, Y.; Xu, B. Ts2vec: Towards universal representation of time series. In Proceedings of the AAAI Conference on Artificial Intelligence, Online, 22 Februry–1 March 2022; pp. 8980–8987.
13. Mehta, S.; Rastegari, M.; Caspi, A.; Shapiro, L.; Hajishirzi, H. Espnet: Efficient spatial pyramid of dilated convolutions for semantic segmentation. In Proceedings of the European Conference on Computer Vision (ECCV), Munich, Germany, 8–14 September 2018; pp. 552–568.
14. Qiu, X.; Cheng, H.; Wu, X.; Hu, J.; Guo, C.; Yang, B. A comprehensive survey of deep learning for multivariate time series forecasting: A channel strategy perspective. *arXiv* **2025**, arXiv:2502.10721.

15. Sørensen, M.L.; Nystrup, P.; Bjerregård, M.B.; Møller, J.K.; Bacher, P.; Madsen, H. Recent developments in multivariate wind and solar power forecasting. *Wiley Interdiscip. Rev. Energy Environ.* **2023**, *12*, e465. [CrossRef]
16. Nie, Y.; Nguyen, N.H.; Sinthong, P.; Kalagnanam, J. A time series is worth 64 words: Long-term forecasting with transformers. *arXiv* **2022**, arXiv:2211.14730.
17. Zeng, A.; Chen, M.; Zhang, L.; Xu, Q. Are transformers effective for time series forecasting? In Proceedings of the AAAI Conference on Artificial Intelligence, Washington, DC, USA, 7–14 February 2023; pp. 11121–11128. [CrossRef]
18. Lin, S.; Lin, W.; Hu, X.; Wu, W.; Mo, R.; Zhong, H. Cyclenet: Enhancing time series forecasting through modeling periodic patterns. *Adv. Neural Inf. Process. Syst.* **2024**, *37*, 106315–106345.
19. Han, L.; Ye, H.-J.; Zhan, D.-C. The capacity and robustness trade-off: Revisiting the channel independent strategy for multivariate time series forecasting. *IEEE Trans. Knowl. Data Eng.* **2024**, *36*, 7129–7142. [CrossRef]
20. Chang, C.; Wang, W.-Y.; Peng, W.-C.; Chen, T.-F. Llm4ts: Aligning pre-trained llms as data-efficient time-series forecasters. *ACM Trans. Intell. Syst. Technol.* **2025**, *16*, 1–20. [CrossRef]
21. Jin, M.; Wang, S.; Ma, L.; Chu, Z.; Zhang, J.Y.; Shi, X.; Chen, P.-Y.; Liang, Y.; Li, Y.-F.; Pan, S. Time-llm: Time series forecasting by reprogramming large language models. *arXiv* **2023**, arXiv:2310.01728.
22. Ansari, A.F.; Stella, L.; Turkmen, C.; Zhang, X.; Mercado, P.; Shen, H.; Shchur, O.; Rangapuram, S.S.; Arango, S.P.; Kapoor, S. Chronos: Learning the language of time series. *arXiv* **2024**, arXiv:2403.07815.
23. De Caro, F.; De Stefani, J.; Vaccaro, A.; Bontempi, G. DAFT-E: Feature-based multivariate and multi-step-ahead wind power forecasting. *IEEE Trans. Sustain. Energy* **2021**, *13*, 1199–1209. [CrossRef]
24. Chen, S.-A.; Li, C.-L.; Yoder, N.; Arik, S.O.; Pfister, T. Tsmixer: An all-mlp architecture for time series forecasting. *arXiv* **2023**, arXiv:2303.06053.
25. Zheng, X.; Chen, X.; Schürch, M.; Mollaysa, A.; Allam, A.; Krauthammer, M. Simts: Rethinking contrastive representation learning for time series forecasting. *arXiv* **2023**, arXiv:2303.18205.
26. Yang, L.; Zhang, R.-Y.; Li, L.; Xie, X. Simam: A simple, parameter-free attention module for convolutional neural networks. In Proceedings of the International Conference on Machine Learning, Online, 18–24 July 2021; pp. 11863–11874.
27. Shan, C.; Yuan, Z.; Qiu, Z.; He, Z.; An, P. A Dual Arrhythmia Classification Algorithm Based on Deep Learning and Attention Mechanism Incorporating Morphological-temporal Information. In Proceedings of the 2022 International Joint Conference on Neural Networks (IJCNN), Padua, Italy, 18–23 July 2022; pp. 1–8.
28. Liu, F.T.; Ting, K.M.; Zhou, Z.-H. Isolation forest. In Proceedings of the 2008 Eighth IEEE International Conference on Data Mining, Pisa, Italy, 15–19 December 2008; pp. 413–422.
29. Blu, T.; Thévenaz, P.; Unser, M. Linear interpolation revitalized. *IEEE Trans. Image Process.* **2004**, *13*, 710–719. [CrossRef] [PubMed]
30. Ahsan, M.M.; Mahmud, M.P.; Saha, P.K.; Gupta, K.D.; Siddique, Z. Effect of data scaling methods on machine learning algorithms and model performance. *Technologies* **2021**, *9*, 52. [CrossRef]
31. Greff, K.; Srivastava, R.K.; Koutník, J.; Steunebrink, B.R.; Schmidhuber, J. LSTM: A search space odyssey. *IEEE Trans. Neural Netw. Learn. Syst.* **2016**, *28*, 2222–2232. [CrossRef]
32. Cerqueira, V.; Torgo, L.; Mozetič, I. Evaluating time series forecasting models: An empirical study on performance estimation methods. *Mach. Learn.* **2020**, *109*, 1997–2028. [CrossRef]
33. Abnar, S.; Shah, H.; Busbridge, D.; Ali, A.M.E.; Susskind, J.; Thilak, V. Parameters vs flops: Scaling laws for optimal sparsity for mixture-of-experts language models. *arXiv* **2025**, arXiv:2501.12370.
34. Popescu, M.-C.; Balas, V.E.; Perescu-Popescu, L.; Mastorakis, N. Multilayer perceptron and neural networks. *WSEAS Trans. Circuits Syst.* **2009**, *8*, 579–588.
35. Medsker, L.R.; Jain, L. Recurrent neural networks. *Des. Appl.* **2001**, *5*, 2.
36. Oreshkin, B.N.; Carpov, D.; Chapados, N.; Bengio, Y. N-BEATS: Neural basis expansion analysis for interpretable time series forecasting. *arXiv* **2019**, arXiv:1905.10437.
37. Zhang, K.; Sun, M.; Han, T.X.; Yuan, X.; Guo, L.; Liu, T. Residual networks of residual networks: Multilevel residual networks. *IEEE Trans. Circuits Syst. Video Technol.* **2017**, *28*, 1303–1314. [CrossRef]
38. Hollis, T.; Viscardi, A.; Yi, S.E. A comparison of LSTMs and attention mechanisms for forecasting financial time series. *arXiv* **2018**, arXiv:1812.07699.
39. Brauwers, G.; Frasincar, F. A general survey on attention mechanisms in deep learning. *IEEE Trans. Knowl. Data Eng.* **2021**, *35*, 3279–3298. [CrossRef]
40. Lin, S.; Lin, W.; Wu, W.; Chen, H.; Yang, J. Sparsetsf: Modeling long-term time series forecasting with 1k parameters. *arXiv* **2024**, arXiv:2405.00946.

Disclaimer/Publisher's Note: The statements, opinions and data contained in all publications are solely those of the individual author(s) and contributor(s) and not of MDPI and/or the editor(s). MDPI and/or the editor(s) disclaim responsibility for any injury to people or property resulting from any ideas, methods, instructions or products referred to in the content.

energies

MDPI

Article

Research on 10 kV Cable Insulation Detection Method Based on Ground Current Phase Variation

Gang Liu [1], Yuanming Zhang [1], Tonghui Ye [2], Dongdong Zhang [2,*], Peigen Cao [1] and Yulan Che [1]

1 State Grid Suqian Power Supply Company, Suqian 223800, China; masonlf586@gmail.com (G.L.); 04121607love@gmail.com (Y.Z.); jsidiwne@gmail.com (P.C.); sunnycoal709@gmail.com (Y.C.)
2 School of Electric Power Engineering, Nanjing Institute of Technology, Nanjing 211167, China; y00450230642@njit.edu.cn
* Correspondence: zhangdd@njit.edu.cn

Abstract

In view of the limitations of traditional offline detection and external excitation online detection of 10 kV cables, this paper proposes a method to evaluate the insulation aging condition of power cables by online measuring of the phase angle of the cable's ground current, and explores the impact of load fluctuations on cable insulation. By setting the relative permittivity of the cable to characterize the phase variation of the ground current under different aging degrees, and analyzing the phase variation of the cable's ground current under different load changes at the same aging degree, a load correction-based dynamic dielectric loss evaluation method for cables is proposed. Through the construction of cable simulation models and the processing of field data, the following conclusions have been reached: Under a 1 MW load, the phase angle of the sheath grounding current in the aged phase increases as the dielectric constant of the insulation increases. At the same aging degree, with an increase in load, the phase differences of the aging phase sheath ground current and the steel armor ground current both show a decreasing trend. To eliminate the impact of load, a dynamic dielectric loss load correction method is proposed, and combined with field data analysis, the dynamic dielectric loss of cables under different loads is corrected to a 1 MW load. Specifically: Under 0.3 MW, the correction coefficients k for the sheath and steel armor are 0.609 and 0.778, respectively. Under 3.5 MW, the correction coefficients k for the sheath and steel armor are 1.435 and 1.089, respectively. This study provides a theoretical basis and experimental verification for online cable monitoring methods.

Keywords: cable aging; online detection; grounding current phase; load; dynamic dielectric loss

1. Introduction

Due to the rapid development of urbanization, underground cables play an increasingly vital role in urban power supply and are widely used [1]. However, power cables have inherent drawbacks, such as unstable insulation performance during manufacturing, potential insulation damage due to collisions during installation, and long-term exposure to humid underground environments that can lead to water tree aging [2,3], as shown in Figure 1. Statistics show that 32% of urban distribution cables in China have been in operation for over 15 years, and insulation aging accounts for more than 47% of total cable failures. These issues cause insulation degradation, significantly reducing performance and leading to insulation breakdown accidents [4]. It can be seen that the aging of cables has a non-negligible impact on the normal operation of the power grid. Effective fault detection of cables should be carried out as soon as possible to prevent serious losses.

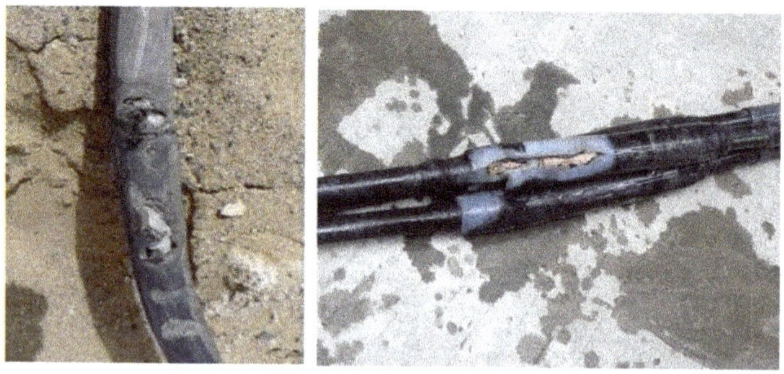

Figure 1. Insulation damage caused by overheating and moisture.

Extensive research has been conducted on cable insulation aging detection [5–8]. For example, reference [9] studied water tree and thermal aging in cable insulation, simulating aging experiments on power cable samples and determining aging levels by comparing $tan\delta$-φ curves under different aging conditions. In reference [10], tensile, breakdown strength, infrared spectroscopy, and dielectric spectroscopy tests were performed on XLPE cables, establishing relationships between aging time and parameters such as tensile strength, breakdown strength, hydroxyl index, and dielectric loss factor. Reference [11] proposed an oscillating wave partial discharge detection method for high-voltage XLPE cables using LSTM algorithms, which locates defects by analyzing the time differences between the original and reflected waves. Reference [12] provides a detailed introduction to partial discharge detection methods, including the offline detection of AC test, VLF test, and DAC test, as well as online detection methods such as HFCT, VHF, AE, etc., and introduces their principles and applicability. Although these studies achieved results, most of the offline detection methods are destructive, and have the disadvantages of large test equipment and high operation requirements. Moreover, the online detection methods are affected by serious electromagnetic interference and signal attenuation [13–15]. Research on reliable online detection without external excitation remains scarce [16–18]. Based on this, this paper aims to propose an online detection method for the variation of the grounding current phase angle of power cables with aging. This method does not require external excitation and compares the phase angle relationships of cable grounding currents under different degrees of simulated aging through simulations and experiments, in order to determine the aging status of the cable.

However, in practical engineering applications, changes in the load side of cable lines can have a certain degree of impact on cable insulation. When the load side equipment runs for a long time, a large amount of heat will be generated and conducted to the cable, causing a sharp increase in cable temperature. High temperature conditions can accelerate the thermal aging process of cable insulation materials, reducing their tensile and bending strength and making them more susceptible to mechanical damage and insulation breakdown [19–21]. And when the load size changes, it will directly cause a change in the current size of the cable, thereby causing a change in the phase angle of the cable's ground current. Therefore, this article establishes aging models for power cables under different loads, analyzes the influence of load on the phase angle of the ground current with aging, and proposes a method for dynamic dielectric loss load correction. By combining field data collection with analysis, the dielectric loss measurements under different load conditions are normalized to a 1 MW reference, thereby mitigating the influence of load variations on detection accuracy. The results can provide a reference for online detection of aging in power cables.

2. Phase Analysis Model for 10 kV Three-Core Pipe-Type Cables

This section builds a 10 kV three-phase cable model based on simulation software, and simulates cable aging by changing the relative dielectric constant of its insulation layer.

2.1. Cable Simulation Model

The 10 kV cable simulation model, shown in Figure 2, consists of a power supply side, input side, and load side. The power supply uses a 110 kV three-phase ideal source, stepped down to a 10 kV equivalent source. The output connects to a 1 MW load to simulate actual usage, while the input side includes current transformers to capture shielding layer and armor grounding currents.

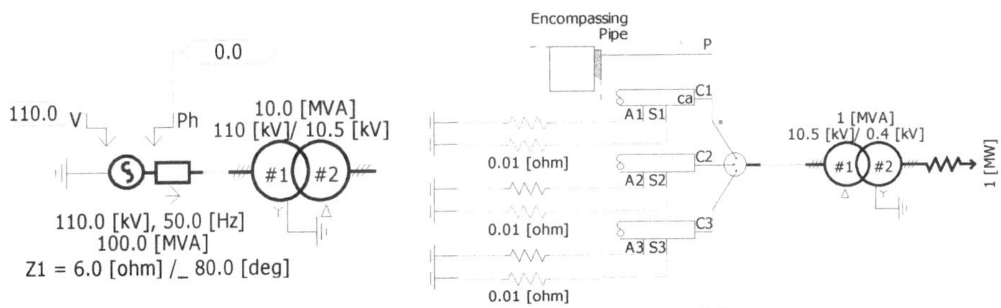

(a) (b)

Figure 2. Simulation model of 10 kV distribution network power cable: (**a**) power supply side; (**b**) load side.

2.2. Cable Structure and Aging Simulation

The three-phase cable arrangement and single-cable structure are depicted in Figure 3. A single cable comprises a conductor, inner semiconductor layer, insulation layer, outer semiconductor layer, metal shielding layer, filler layer, inner sheath, armor layer, and outer sheath.

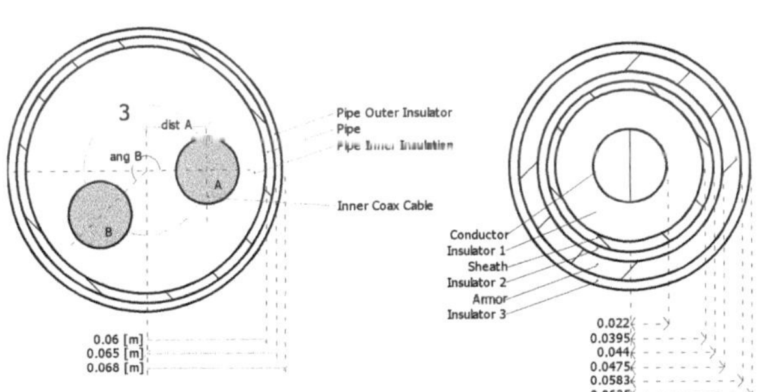

Figure 3. Structure of 10 kV three-core pipe-type cable.

The simulation focuses on phase angle changes in grounding currents after aging, considering non-uniform single-phase aging. Aging is simulated by adjusting the relative permittivity of the insulation layer: the non-aged phases are set to 2.0, while the aged phase is set to 2.0, 2.4, 2.8, 3.2, and 5.0 to observe phase angle variations in the shielding and armor grounding currents [15].

3. Analysis of Grounding Current Phase Angle Under Cable Aging

In the study of the cable to ground current phase angle, by analyzing the changes in the angle's amplitude, the aging condition and specific aging phase of cable insulation can be determined. Therefore, in this section, which covers research on the impact of load side on cable insulation, we choose to compare and analyze the changes in the cable to ground current phase angle under single-phase non-uniform aging conditions to verify the sensitivity of the grounding current phase angle to cable insulation aging characteristics and the accuracy of determining the cable insulation's aging phase.

3.1. Phase Angle Analysis Under 1 MW Load

In simulations using a 1 MW load, the phase angle data for the sheath grounding current of the aged phase (A-phase) is extracted, as shown in Figure 4.

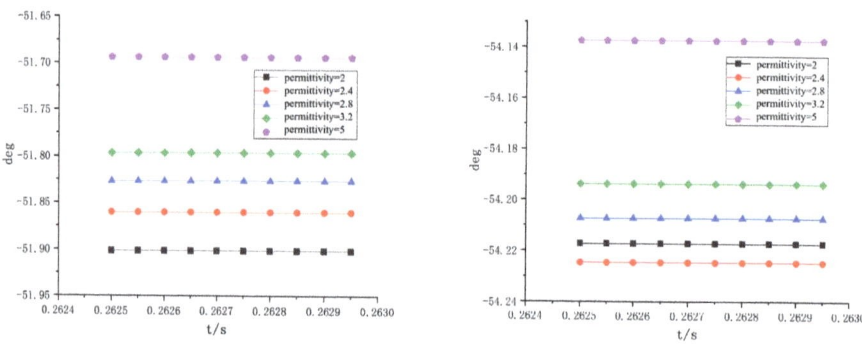

(a) (b)

Figure 4. Grounding current phase angles for aged A-phase: (**a**) sheath grounding current phase angle; (**b**) armor grounding current phase angle.

By observing the phase angle changes in the above figure, it can be seen that as the dielectric constant gradually increases from 2 to 5, the phase angle of the aging A-phase sheath to ground current increases from $-51.9°$ to $-51.69°$, while the phase angle of the steel armor to ground current increases from $-54.21°$ to $-54.13°$. The amplitude changes of the phase angles of the two currents are $0.21°$ and $0.09°$, respectively, and the phase angle of the sheath current shows a gradually increasing trend. Through comparative analysis, it can be clearly observed that the amplitude change of the phase angle of the sheath to ground current is greater. To further investigate the current phase angle changes of the cable, a phase angle graph of the ground current for phase B of the unaged cable was drawn, as shown in Figure 5.

From the above figure, it can be seen that as the dielectric constant increases, the phase angle of the unaged B-phase sheath to ground current increases from $-171.9°$ to $-171.85°$, while the phase angle of the steel armor to ground current increases from $-174.217°$ to $-174.211°$. The amplitude changes of the current phase angles of the two are $0.05°$ and $0.006°$, respectively. It can be seen that the amplitude changes of the phase angle of the unaged B-phase are relatively small, and the trends of the phase angle changes of the sheath and steel armor currents are inconsistent.

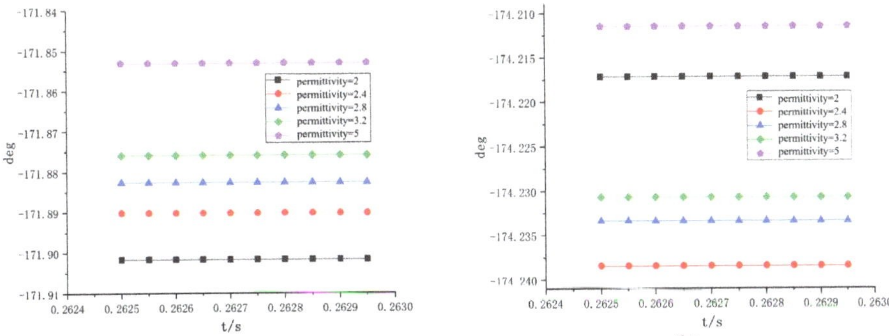

(**a**) (**b**)

Figure 5. Grounding current phase angles for non-aged B-phase: (**a**) sheath grounding current phase angle; (**b**) armor grounding current phase angle.

3.2. Phase Angle Analysis Under 2 MW Load

Analogous to the above analysis method, the parameters on the load side are simulated to be 2 MW. By simulating the uneven aging state of the cable, the current phase angle characteristics of cable aging phase A are analyzed, as shown in Figure 6.

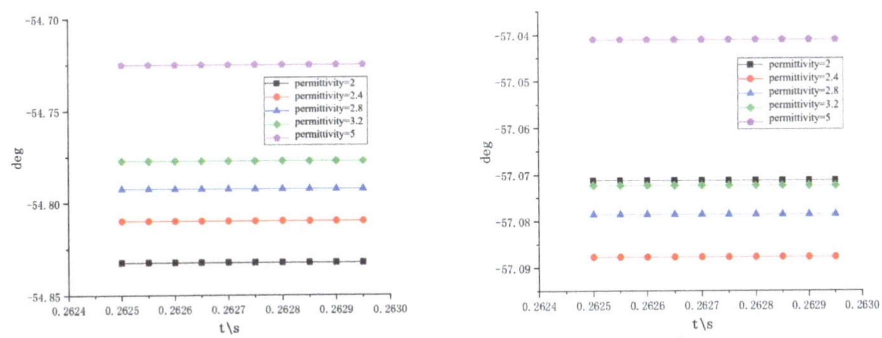

(**a**) (**b**)

Figure 6. Grounding current phase angles for aged A-phase: (**a**) sheath grounding current phase angle; (**b**) armor grounding current phase angle.

After comparing the above subfigures, it can be seen that as the dielectric constant increases, the phase angle of the cable sheath to ground current increases from −54.83° to −54.72°, while the phase angle of the steel armor to ground current increases from −57.07° to −57.04°. The increases in the phase angle amplitude of the two are 0.11° and 0.03°, respectively. By analyzing the trend of the current phase angle changes in aging phase A, it can be seen that the sheath to ground current phase angle shows a gradually increasing trend, while the changes in the steel armor to ground current phase angle are relatively irregular, and the two trends do not reflect consistency. In order to compare the phase angle variation characteristics of the aged and unaged phases of cables more specifically, a current phase angle variation diagram of the unaged phase B was drawn, as shown in Figure 7.

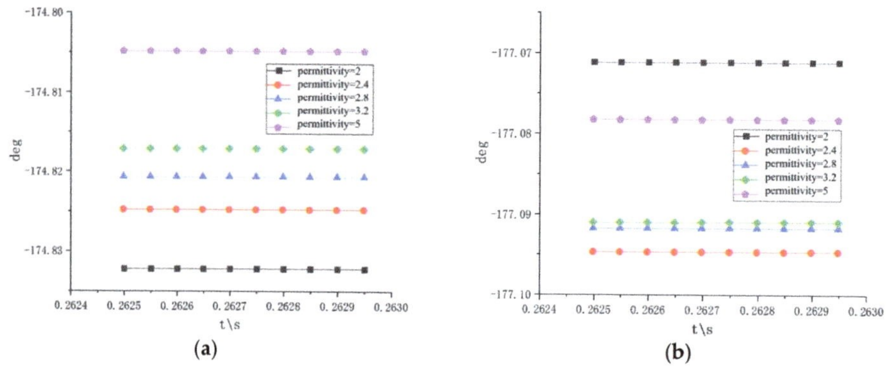

Figure 7. Grounding current phase angles for aged A-phase: (**a**) sheath grounding current phase angle; (**b**) armor grounding current phase angle.

By observing the amplitude changes of the current phase angle during the cable's B phase, it can be seen that as the dielectric constant increases, the phase angle of the sheath to ground current increases from −174.83° to −174.80°, while the phase angle of the steel armor to ground current decreases from −177.07° to −177.08°. The amplitude changes of the two phase angles are 0.03° and −0.01°, respectively. From the data in the above figure, it can be seen that the phase angle change during the unaged B-phase is relatively small, and there is no correlation between the phase angle change characteristics of the sheath and steel armor currents.

3.3. Phase Angle Analysis Under Different Loads

As can be seen from the previous text, when the cable ages, the phase angle of the grounding current of the cable shows a regular change as the degree of aging increases. Different cable loads can have a certain impact on this pattern, such as changes in the phase angle difference. In order to explore the influence of different loads on the online monitoring of the aging status of the grounding current phase angle of the cable, this section sets four different load side parameters of 0.5 MW, 1 MW, 1.5 MW, and 2 MW for cables under four different working conditions, and compares the phase angle change patterns of the cables under aging compared to normal conditions.

We simulated single-phase aging conditions under four different loads, with a dielectric constant of 2.8 for the aging cable. The phase difference between the grounding currents under normal conditions is shown in Figure 8.

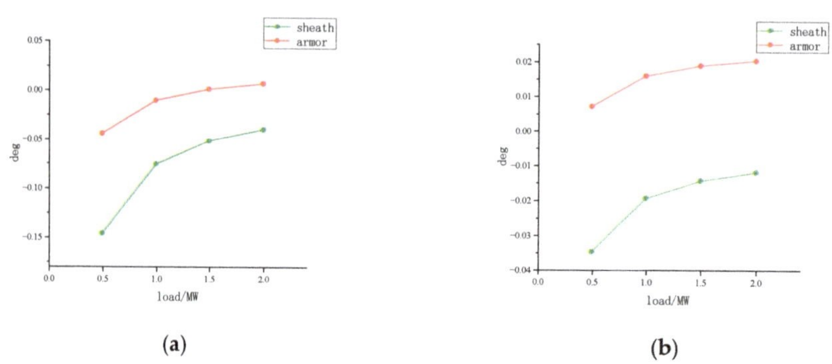

Figure 8. Grounding current phase differences under different loads: (**a**) aged phase cable; (**b**) non-aged phase cable.

As the load increases, the phase difference for the aged phase decreases: the sheath phase difference is reduced from −0.145° to −0.039°, and the armor phase difference decreases from −0.044° to 0.007°. For the non-aged phase, the sheath phase difference decreases from −0.035° to −0.011°, while the armor phase difference increases from 0.007° to 0.02°. The sheath phase difference is most affected by load changes, indicating that load variations can influence the accuracy of online aging detection.

4. Dynamic Dielectric Loss Correction Method Under Different Loads

To mitigate load interference, a dynamic dielectric loss correction method is proposed.

4.1. Dynamic Dielectric Loss Calculation

A vector diagram of three-phase cable currents is shown in Figure 9.

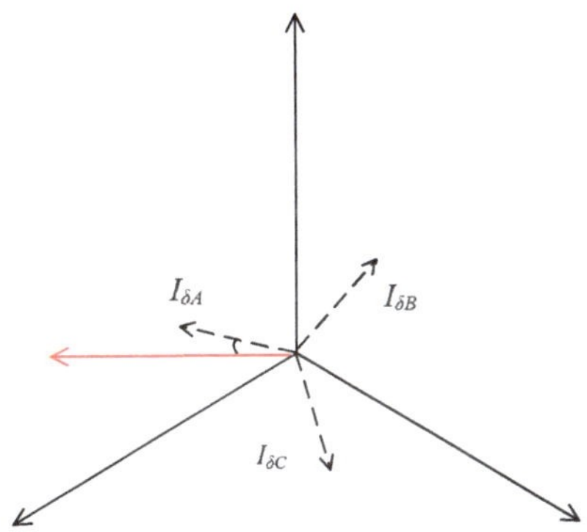

Figure 9. Cable current phase vector diagram.

Here, I_A, I_B, and I_C are the main currents, while $I_{\delta A}$, $I_{\delta B}$, and $I_{\delta C}$ are the grounding currents. The dynamic dielectric loss angle δ is defined as the complement of the angle between the grounding current and the main current:

$$\delta_A \approx 90^\circ - \angle(I_{\delta A}, I_A). \quad (1)$$

Under normal cable conditions, distributed capacitance is formed between the conductor of the cable and the insulation and shielding layers. When the insulation is fine, the resistive component of the leakage current is extremely small, while the capacitive component dominates. Therefore, the phase difference between the grounding current and the main current is about 90°. When the cable ages, the insulation resistance decreases, the resistive component of the leakage (grounding) current increases, and the grounding current changes from being mainly capacitive to exhibiting a mixture of resistance and capacitance, resulting in a decrease in the phase difference between the grounding current and the main current, that is, an increase in the dynamic dielectric loss angle δ_A and an increase in the dynamic dielectric loss $tan\delta_A$. When the cable load increases, the main current increases and the cable heats up, causing changes in insulation resistance and also causing changes in δ_A. Therefore, this section studies the method of dynamic dielectric loss discrimination for insulation aging based on load correction, and summarizes its variation pattern through analysis of field data.

4.2. Cable Current Signal Collection and Analysis

The signal of the grounding current of the cable box during the operation of a branch line is collected by using clamp-type transformers and oscilloscopes. A diagram of the current field collection model and a photo of it in the field are shown in Figure 10. The 10 kV branch coaxial cable is transmitted through underground cables, and each phase cable is stripped and led out to the ring-shaped main unit. The model used is ZR-YJV22-8.7/15-3X400. By using current transformers to collect the main current of the cable, the grounding current of the steel cable and the grounding current of the external shield can be used to obtain the current signals.

(a) (b)

Figure 10. Underground cable model and field photo: (**a**) current collection model; (**b**) field collection.

The test cable parameters are shown in Table 1. The experiment used three cables with the same operating years under the same branch line. According to the on-site introduction, the F70 cable was used for residential electricity, the F71 cable distributed energy to commercial stores, and the F72 cable was used to supply power to roadside lighting lamps. Due to the on-site time period and relevant personnel, the approximate load situation was understood, and the copper shielding (sheath) grounding current and armor grounding current signals were collected separately.

Table 1. Test cable parameters.

Test Cable	Cable Switch	Approximate Load
10 kV branch line	F70	3.5 MW
	F71	1 MW
	F72	0.3 MW

The phase and dynamic dielectric loss angles of the current signal collected and processed by the oscilloscope are shown in Table 2.

Table 2. Collected phase and dielectric loss angles.

Cable Switch	Phase	Sheath Phase Angle	Armor Phase Angle	Main Current Angle	Dynamic Dielectric Loss Angle δ_1 of Sheath to Main Current Phase Difference	Dynamic Dielectric Loss Angle δ_2 of Armor to Main Current Phase Difference
F70	A	24.6°	23.7°	−63.5°	1.9°	2.8°
F71	A	−80.3°	−81.6°	−169°	1.3°	2.6°
F72	A	51.4°	50.2°	−37.8°	−0.8°	2°

Based on the above dielectric loss angle calculation method, the dynamic dielectric loss under different loads is normalized and corrected according to a 1 MW load. The correction coefficient k is as follows:

$$k = \frac{tan\delta}{tan\delta_0}. \quad (2)$$

In the formula, δ represents the dynamic dielectric loss angle under different loads, and δ_0 represents the dynamic dielectric loss angle under a load of 1 MW. The dynamic dielectric loss and the correction factor are obtained as shown in Table 3.

Table 3. Dynamic dielectric loss and correction coefficients.

Load	$tan\delta_1$	$tan\delta_2$	Correction Coefficient k_1	Correction Coefficient k_2
3.5 MW	0.033	0.049	1.435	1.089
1 MW	0.023	0.045	1	1
0.3 MW	0.014	0.035	0.609	0.778

In summary, by correcting the dynamic dielectric loss under different loads, normalized dynamic dielectric loss values can be obtained to evaluate cable aging and eliminate the influence of load on signal acquisition.

5. Conclusions

This paper investigates the phase angle changes of grounding currents in 10 kV three-core cables under aging and varying loads, analyzes the impact of load on aging assessment, and proposes a dynamic dielectric loss correction method. The key conclusions are as follows:

1. For single non-uniform aging, the phase of the grounding current of the aging phase sheath shows an increasing trend with increases in the dielectric constant, and the change amplitude is significantly greater than that of the non-aging phase, which can indicate a change in the phase of the aging state.
2. Observing the simulation of cable aging under 1 MW and 2 MW loads, the phase angle of the grounding current of the steel cable first decreases and then increases with the deepening of cable aging, and there is no overall regularity. However, the amplitude of the phase angle change of the grounding current during the aging phase of the steel cable is greater than that of the non-aging phase, which can be used to distinguish aged cables.
3. When the degree of cable aging is the same, the phase angle changes of the cable grounding current are different under different loads. As the cable load increases, the phase difference of the aging phase grounding current shows a gradually decreasing trend. The phase difference of the sheath grounding current decreases from $-0.145°$ to $-0.039°$, the phase difference of the armor grounding current decreases from $-0.044°$ to $0.007°$, the phase difference of the non-aging phase sheath grounding current decreases from $-0.035°$ to $-0.011°$, and the phase difference of the armor grounding current increases from $0.007°$ to $0.02°$. Therefore, it can be seen that the load size of the cable will, to some extent, affect the accuracy of online monitoring of cable aging based on the phase difference of the grounding current.
4. By defining the dynamic dielectric loss angle δ, a correction coefficient k based on load normalization is proposed. Experimental data shows that under different loads, the dynamic dielectric loss $tan\delta_1$ needs to be normalized to a 1 MW benchmark through

the correction coefficient k to eliminate load interference and improve the accuracy of aging assessment under different load conditions.

The phase change of the grounding current can, to some extent, characterize cable aging, of which the phase difference of the sheath current is a key indicator, but it needs to be corrected based on actual load conditions to improve accuracy. This research provides a theoretical basis and experimental verification for online cable detection methods.

Author Contributions: Conceptualization, G.L. and T.Y.; methodology, Y.Z., D.Z., and T.Y.; software, G.L., Y.Z., and P.C.; validation, G.L. and T.Y.; formal analysis, D.Z. and Y.C.; investigation, P.C. and Y.C.; resources, G.L.; data curation, T.Y.; writing—original draft preparation, G.L.; writing—review and editing, Y.Z.; supervision, G.L. and T.Y.; project administration, D.Z.; funding acquisition, D.Z. All authors have read and agreed to the published version of the manuscript.

Funding: This research was funded by the Science Foundation of State Grid Jiangsu Electric Power Company (J2023101).

Data Availability Statement: The original contributions presented in the study are included in the article; further inquiries can be directed to the corresponding author.

Acknowledgments: The authors would like to express their gratitude to Nanjing Institute of Technology for their assistance with this research.

Conflicts of Interest: Authors Gang Liu, Yuanming Zhang, Peigen Cao and Yulan Che were employed by the company State Grid Suqian Power Supply Company. The remaining authors declare that the research was conducted in the absence of any commercial or financial relationships that could be construed as a potential conflict of interest.

References

1. Li, C.Y.; Tan, X.; Liu, L.G.; Zhang, W.; Yang, Q.M.; Cao, J.Y.; Zhou, E.C.; Li, M.Z.; Song, Z.X. Investigation of Near-Infrared Spectroscopy for Assessing the Macroscopic Mechanical Properties of Cross-Linked Polyethylene During Thermal Aging. *Materials* **2025**, *18*, 504. [CrossRef] [PubMed]
2. Afia, R.S.A.; Mustafa, E.; Tamus, Z.A. Comparison of Mechanical and Low-Frequency Dielectric Properties of Thermally and Thermo-Mechanically Aged Low Voltage CSPE/XLPE Nuclear Power Plant Cables. *Electronics* **2021**, *10*, 2728. [CrossRef]
3. Zhu, X.; Yin, Y.; Wu, J.D.; Wang, X.F. Study on Aging Characteristics of XLPE Cable Insulation Based on Quantum Chemical Calculation. *IEEE Trans. Dielectr. Electr. Insul.* **2020**, *27*, 1942–1950. [CrossRef]
4. Yang, J.; Li, R.Z.; Hu, L.Y.; Wang, W.W. Influence of Thermal Aging on Space Charge Characteristics and Electrical Conduction Behavior of Cross-Linked Polyethylene Cable Insulation. *Polymers* **2024**, *16*, 1600. [CrossRef] [PubMed]
5. Cheng, D.F.; Wei, J.H.; Huang, F.B.; Wu, Q.; Tao, X.F.; Wu, K.; Li, Q.; Xue, J.Y. Identification of local thermal aging defects in power cables based on frequency domain reflection coefficient spectroscopy method. *J. Phys. Conf. Ser.* **2024**, *2917*, 012007. [CrossRef]
6. Wang, R.; Pan, R. A Cable Defect Assessment Method Based on a Mixed-Domain Multi-Feature Set of Overall Harmonic Signals. *Energies* **2025**, *18*, 83. [CrossRef]
7. Zhou, K.; Yuan, H.; Li, Y.; Li, M.Z.; Li, Z.R.; Lin, S.Y. Assessing Aging Status and Type of XLPE Cable Insulation With a Graphic Approach Based on PDC Measurement. *Ieee Trans. Power Deliv.* **2022**, *37*, 5114–5123. [CrossRef]
8. Jahromi, A.N.; Pattabi, P.; Densley, J.; Lamarre, L. Medium voltage XLPE cable condition assessment using frequency domain spectroscopy. *Ieee Electr. Insul. Mag.* **2020**, *36*, 9–18. [CrossRef]
9. Kim, C.; Jin, Z.; Huang, X.; Jiang, P.; Ke, Q. Investigation on water treeing behaviors of thermally aged XLPE cable insulation. *Polym. Degrad. Stab.* **2007**, *92*, 537–544. [CrossRef]
10. Ma, F.; Zhang, C. Diagnostic Methods for Insulation Aging of Cross-Linked Polyethylene Cables. *Beijing Polytech. Vocat. Coll.* **2022**, *21*, 11–16.
11. Han, X. Oscillating Wave Partial Discharge Detection Method for High-Voltage Cross-Linked Cable Lines Based on the LSTM Algorithm. *Electr. Mach. Control Appl.* **2022**, *49*, 41–46+80.
12. Suganya, G.; Adolfo, M.; Alfredo, J.A.; Narasimman, P. A review on partial discharge diagnosis in cables: Theory, techniques, and trends. *Measurement* **2023**, *216*, 112882.

13. Fu, H.; Qiu, L.; Ai, Y.H.; Tu, J.; Yan, Y.T. Deep learning-based fault detection and location in underground power cables using resonance frequency analysis. *Electr. Eng.* **2024**, *107*, 1–12. [CrossRef]
14. Nilsson, D.J.; Gubanski, S.M.; Serdyuk, Y.V. Electrical Detection of Degradation in Specimens of HVDC Cable Insulation. *Energies* **2020**, *13*, 3963. [CrossRef]
15. Chen, C.; Cheng, C.H.; Wang, X.; Wu, K. Space Charge Characteristics for XLPE Coaxial Cable Insulation Under Electrothermal Accelerated Aging. *IEEE Trans. Dielectr. Electr. Insul.* **2022**, *29*, 727–736.
16. Wang, Y.; Zhao, A.; Zhang, X.; Shen, Y.; Yang, F.; Deng, J.; Zhang, G. Study of Dielectric Response Characteristics for Thermal Aging of XLPE Cable Insulation. In Proceedings of the International Conference on Condition Monitoring and Diagnosis (CMD), Xian, China, 25–28 September 2016; pp. 602–605.
17. Chen, H.; Wang, J.J.; Hu, H.J.; Li, X.F.; Huang, Y.Y. Aging Detection of 110 kV XLPE Cable for a CFETR Power Supply System Based on Deep Neural Network. *Energies* **2022**, *15*, 3127. [CrossRef]
18. Wang, X.; Liu, E.H.; Zhang, B. Reflectometry-Based Cable Insulation Aging Diagnosis and Prognosis. *IEEE Trans. Ind. Electron.* **2022**, *69*, 4148–4157. [CrossRef]
19. Liu, Y.; Cao, X.L. Insulation Performance Evaluation of HV AC/DC XLPE Cables by 0.1 Hz Tan δ Test on Circumferentially Peeled Samples. *IEEE Trans. Dielectr. Electr. Insul.* **2017**, *24*, 3941–3950. [CrossRef]
20. Rizwan, T.; Ibrahim, A.; Ur, R.A.; Elsayed, T.E.; Muhammad, S.; Ghamry, N.A.; Habib, H. An Optimized Solution for Fault Detection and Location in Underground Cables Based on Traveling Waves. *Energies* **2022**, *15*, 6468. [CrossRef]
21. Sebbane, Y.; Boubakeur, A.; Mekhaledi, A. Influence of Thermal Aging and Water Adsorption on XLPE Cables Insulation Mechanical and Physico-chemical Properties. *IEEE Trans. Dielectr. Electr. Insul.* **2021**, *28*, 1694–1702. [CrossRef]

Disclaimer/Publisher's Note: The statements, opinions and data contained in all publications are solely those of the individual author(s) and contributor(s) and not of MDPI and/or the editor(s). MDPI and/or the editor(s) disclaim responsibility for any injury to people or property resulting from any ideas, methods, instructions or products referred to in the content.

MDPI AG
Grosspeteranlage 5
4052 Basel
Switzerland
Tel.: +41 61 683 77 34

Energies Editorial Office
E-mail: energies@mdpi.com
www.mdpi.com/journal/energies

Disclaimer/Publisher's Note: The title and front matter of this reprint are at the discretion of the Guest Editors. The publisher is not responsible for their content or any associated concerns. The statements, opinions and data contained in all individual articles are solely those of the individual Editors and contributors and not of MDPI. MDPI disclaims responsibility for any injury to people or property resulting from any ideas, methods, instructions or products referred to in the content.